Communications in Computer and Information Science

1746

Rationale

The CCIS series is devoted to the publication of proceedings of computer science conferences. Its aim is to efficiently disseminate original research results in informatics in printed and electronic form. While the focus is on publication of peer-reviewed full papers presenting mature work, inclusion of reviewed short papers reporting on work in progress is welcome, too. Besides globally relevant meetings with internationally representative program committees guaranteeing a strict peer-reviewing and paper selection process, conferences run by societies or of high regional or national relevance are also considered for publication.

Topics

The topical scope of CCIS spans the entire spectrum of informatics ranging from foundational topics in the theory of computing to information and communications science and technology and a broad variety of interdisciplinary application fields.

Information for Volume Editors and Authors

Publication in CCIS is free of charge. No royalties are paid, however, we offer registered conference participants temporary free access to the online version of the conference proceedings on SpringerLink (http://link.springer.com) by means of an http referrer from the conference website and/or a number of complimentary printed copies, as specified in the official acceptance email of the event.

CCIS proceedings can be published in time for distribution at conferences or as post-proceedings, and delivered in the form of printed books and/or electronically as USBs and/or e-content licenses for accessing proceedings at SpringerLink. Furthermore, CCIS proceedings are included in the CCIS electronic book series hosted in the SpringerLink digital library at http://link.springer.com/bookseries/7899. Conferences publishing in CCIS are allowed to use Online Conference Service (OCS) for managing the whole proceedings lifecycle (from submission and reviewing to preparing for publication) free of charge.

Publication process

The language of publication is exclusively English. Authors publishing in CCIS have to sign the Springer CCIS copyright transfer form, however, they are free to use their material published in CCIS for substantially changed, more elaborate subsequent publications elsewhere. For the preparation of the camera-ready papers/files, authors have to strictly adhere to the Springer CCIS Authors' Instructions and are strongly encouraged to use the CCIS LaTeX style files or templates.

Abstracting/Indexing

CCIS is abstracted/indexed in DBLP, Google Scholar, EI-Compendex, Mathematical Reviews, SCImago, Scopus. CCIS volumes are also submitted for the inclusion in ISI Proceedings.

How to start

To start the evaluation of your proposal for inclusion in the CCIS series, please send an e-mail to ccis@springer.com.

Alvaro David Orjuela-Cañón · Jesus Lopez ·
Julian David Arias-Londoño ·
Juan Carlos Figueroa-García
Editors

Applications of Computational Intelligence

5th IEEE Colombian Conference, ColCACI 2022
Cali, Colombia, July 27–29, 2022
Revised Selected Papers

 Springer

Editors
Alvaro David Orjuela-Cañón ⓘD
Universidad del Rosario
Bogotá, Colombia

Julian David Arias-Londoño ⓘD
Universidad Politécnica de Madrid
Madrid, Spain

Jesus Lopez ⓘD
Universidad Autónoma de Occidente
Cali, Colombia

Juan Carlos Figueroa-García ⓘD
Universidad Distrital Francisco
José de Caldas
Bogotá, Colombia

ISSN 1865-0929   ISSN 1865-0937 (electronic)
Communications in Computer and Information Science
ISBN 978-3-031-29782-3   ISBN 978-3-031-29783-0 (eBook)
https://doi.org/10.1007/978-3-031-29783-0

This Springer imprint is published by the registered company Springer Nature Switzerland AG
The registered company address is: Gewerbestrasse 11, 6330 Cham, Switzerland

Preface

The computational intelligence (CI) area is increasingly employed in engineering problems in the Latin America (LA) region. LA scientists have focused their efforts in the CI field as a way to deal with problems of interest for the international community but also of great impact in the LA region. Many different areas, including optimization of energy and transportation systems, computer-aided medical diagnoses, bioinformatics, mining of massive data sets, robotics and automatic surveillance systems, among many others, are commonly addressed problems in this part of the world, because of the great potential those applications could also have in developing countries.

Continuing with the post-pandemic situation due to COVID-19, it was both a challenge and an opportunity to meet again in this edition of the conference. Some works were completely presented in person and others presented by using technological tools. The Colombian Chapter of Computational Intelligence (IEEE CIS) and IEEE Colombia Section chose Cali city, in Colombia, as the venue for the IEEE Colombian Conference on Applications of Computational Intelligence (IEEE ColCACI 2022). Like previous editions, the conference sought to become the most important event in computational intelligence and related fields in Colombia, bringing together attendees from academia, science and industry.

This fifth edition, IEEE ColCACI 2022, was fortified with contributions from scientists, engineers and practitioners working on applications/theory of CI techniques. We received 38 papers by authors from 3 Andean countries, and 23 oral presentations in virtual and in-person mode were carried out. In this way, the conference was an international forum for CI researchers and practitioners to share their most recent advancements and results. The present post-proceedings with revised selected papers include the best 7 papers presented as extended versions of works exhibited at the conference. We will continue working on offering an excellent IEEE ColCACI in future editions.

Finally, we would like to thank the IEEE Colombia Section, the IEEE Computational Intelligence Colombian Chapter, the IEEE Computational Intelligence Society, the Universidad Autónoma de Occidente, the Universidad del Rosario, the Universidad Distrital Francisco José de Caldas, the Universidad de Antioquia, and Springer for their support. Also, special thanks to all volunteers, participants, and the whole crew that worked together to have a successful conference. See you at IEEE ColCACI 2023!

September 2022

Alvaro David Orjuela-Cañón
Julián David Arias-Londoño
Jesus A. Lopez
Juan Carlos Figueroa-García

Organization

General Co-chairs

Jesús Alfonso López Sotelo   Universidad Autónoma de Occidente, Colombia
Alvaro David Orjuela-Cañón   Universidad del Rosario, Colombia

Program Committee Chairs

Julián David Arias-Londoño   Universidad de Antioquia, Colombia
Juan Carlos Figueroa-García   Universidad Distrital Francisco José de Caldas,
    Colombia

Publication Chairs

Alvaro David Orjuela-Cañón   Universidad del Rosario, Colombia
Diana Briceño   Universidad Distrital Francisco José de Caldas,
    Colombia

Financial Chair

José David Cely   Universidad Distrital Francisco José de Caldas,
    Colombia

Webmaster

Fabian Martinez   IEEE Colombia, Colombia

Program Committee

Alvaro David Orjuela-Cañón   Universidad del Rosario, Colombia
Julián David Arias Londoño   Universidad de Antioquia, Colombia
Juan Carlos Figueroa García   Universidad Distrital Francisco José de Caldas,
    Colombia

Carlos Andrés Peña   University of Applied Sciences Western
                                        Switzerland, Switzerland
Edwin Alexander Cerquera   University of Florida, USA
Nadia Nedjah   Universidade Estadual do Río de Janeiro, Brazil
María Daniela López de Luise   CI2S Labs, Argentina
Gustavo Eduardo Juárez   Universidad Nacional de Tucumán, Argentina

Contents

Design of a Segmentation and Classification System for Seed Detection Based on Pixel Intensity Thresholds and Convolutional Neural Networks

Oscar J. Suarez[1]([✉]), Edgar Macias-Garcia[2], Carlos J. Vega[3],
Yersica C. Peñaloza[5], Nicolás Hernández Díaz[4], and Victor M. Garrido[6]

[1] Mechatronics Engineering Department, Universidad de Pamplona,
Pamplona, Colombia
oscar.suarez@unipamplona.edu.co
[2] Intelligent System Research, Intel Labs, Guadalajara, Mexico
[3] School of Management, Universidad del Rosario, Bogota, Colombia
[4] Electrical/Electronic Engineering Department, Universidad Tecnológica de Bolivar,
Cartagena, Colombia
[5] Electronic Engineering Department, Universidad de Pamplona,
Pamplona, Colombia
[6] Electrical Engineering Department, Universidad de Pamplona,
Pamplona, Colombia

Abstract. Due to the computational power and memory of modern computers, computer vision techniques and neural networks can be used to develop a visual inspection system of agricultural products to satisfy product quality requirements. This chapter employs artificial vision techniques to classify seeds in RGB images. As a first step, an algorithm based on pixel intensity threshold is developed to detect and classify a set of different seed types, such as rice, beans, and lentils. Then, the information inferred by this algorithm is exploited to develop a neural network model, which successfully achieves learning classification and detection tasks through a semantic-segmentation scheme. The applicability and satisfactory performance of the proposed algorithms are illustrated by testing with real images, achieving an average accuracy of 92% in the selected set of classes. The experimental results verify that both algorithms can directly detect and classify the proposed set of seeds in input RGB images.

Keywords: Neural Networks · Segmentation · Classification · Pattern Recognition

This work is supported by Universidad de Pamplona, Pamplona, Colombia and Universidad Tecnológica de Bolivar, Cartagena, Colombia.

1   Introduction

Nowadays, there are strict requirements concerning quality control of products due to recent market restrictions, resulting in a demand for increasingly better technologies for more precise testing and better decision-making methods [1,27]. Significant advances in testing techniques and computational technologies have been made to automate production lines to meet the global increase in demand for high-quality and nutrient-dense foods; there is a strong focus on non-invasive and rapid detection methods with the high sensitivity and accuracy required for detecting different seeds and dirt [28]. Additionally, progress in sensor technology has allowed the development of revolutionary techniques for quality control, such as visual inspection systems [6,19,23]. These systems are composed of a device for recording the image, a light source, and a digital signal processor module. In practice, these types of systems improve processing speed while decreasing the cost of inspection processes [7].

Visual inspection systems present several advantages over other techniques, including accuracy and repeatability in hands-free measurements, cost reduction, and elimination of errors associated with the human factor (subjectivity, fatigue, sluggishness, and others). In the branch of agricultural and food products, there is a great diversity of shapes, sizes, colors, and flavors, which are also subdivided into many categories and are allocated to various segments [10,20,22]. A visual inspection requires several processing stages such as image acquisition, feature extraction, and data processing, where computer vision techniques have become an essential tool for ensuring these stages are followed [2,24].

Computer vision techniques are methods that enable computers to infer information from an image and extract distinctive features which are essential for achieving a desired task [27,32]. In the agriculture and food industry, several monitoring goals can be supported by the application of computer vision and pattern recognition techniques [11,21,30], including the extraction of features which directly influence product quality, such as size, shape, texture, brightness, color, and others [3,26]. In this sense, Artificial Neural Networks (NN) play a more significant role in engineering, particularly for identification and classification implicated in computer vision tasks [15,17]. In this topic, deep learning or deep neural networks have been receiving increasing attention due to their accuracy in image processing tasks [3,9,33]. Thanks to an accelerated increase in available storage and computational capabilities, these models have won numerous pattern recognition and machine learning contests in recent years [16,29].

Recently, many works have focused on this application by taking advantage of neural network capacity [4,12,18,25,34]. In [4], a computer image scheme is applied to cluster flax cultivars according to market requirements. In [12], an automated orange classification method is proposed which uses pattern recognition techniques applied to a single-color image of the fruit.

In [34], a novel neural network model for various semantic segmentation tasks is developed. In [25], a state-of-the-art neural network model is presented to localize and classify objects by employing bounding boxes, while in [18], a multi-stage deep learning perception system is developed to achieve different tasks simultaneously. In comparison to the above contributions, the novelty of this work can be summarized as follows:

1. A novel method based on pixel intensity threshold is developed to classify a set of different seed types (rice, beans, and lentils) and dirt.
2. Based on the results achieved by the previous algorithm, a labeled dataset is generated to develop a neural network model using a semantic segmentation scheme, allowing the image pixels to be classified according to the seed types, demonstrating how to relate an analytic algorithm through the generalization capacity of neural networks. Once trained, the model is robust against occlusions and other noise-related types.
3. Finally, a user interface is developed to automatically detect and count different types of seeds in input images.

The rest of the chapter is organized as follows: Sect. 2 discusses a threshold-based segmentation algorithm to detect a proposed set of seeds in input images. Section 3 presents the development of a neural network model designed to accomplish the detection task; first, the threshold-based algorithm is used to generate a dataset composed of RGB images and annotation labels which is used to train the model; then, different model architectures are tested to find the best in terms of accuracy and inference speed. Section 4 covers the development of an user interface to detect and count the different seeds presented in input RGB images using the developed algorithms. Finally, Sect. 5 offers conclusions and future work based on the presented approach.

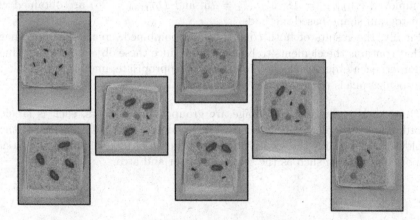

Fig. 1. Some samples of the generated dataset, composed of RGB images including individual and mixed seed types (dirt, rice, beans, and lentils) inside a container.

2   Threshold-Based Segmentation

By considering a defined set of seeds to be detected (rice, beans, and lentils), the first step to achieve a detection scheme must consider the visual features that may allow differentiating the pixels between different classes, including the pixel intensity values on every image color channel and the connection between multiple pixels related to the same object. To detect and classify the different regions in the images according to their seed types, the algorithm presented in Fig. 2 is developed as follows

- In order to generate a dataset, a set of 41 RGB images is generated, where every image includes a container with the different types of seeds involved (Fig. 1).
- Every input RGB image is separated into its red (R) and blue (B) color channels, to calculate the difference between both as a new component.
- A threshold filter is created according to the set's particular pixel values to identify the container area where the seeds are presented. Additionally, noise-removing steps are employed to avoid false positives in subsequent detections.
- Once the container object is delimited, the resulting area is trimmed and decomposed into its red and blue components, obtaining a new image by calculating the difference between these. Thereupon, two conditions ($C_{thresh} = 75$ and $D_{thresh} = 25$) are defined to classify every pixel; the first condition characterizes the objects in the image that present similar color tones (related to the same seed type), while the second characterizes the elements that are not part of the first condition.
- Since the thresholding of the first and second conditions is not 100% effective, it is necessary to remove the noise on a case-by-case basis, either through a range of object size or over a minimum size condition. The thresholds employed ($A_{thresh} = 18$, $C_{thresh} = 75$, and $D_{thresh} = 25$) are obtained via histogram shape-based methods.
- Finally, the results of both conditions are combined, creating a new image that contains the elements to be identified; since these objects have a defined border, each object is cleaned internally as appropriate, making it possible to recognize pixels related to the same object.

Once all the pixels of the image are grouped into objects, each is labeled according to its color and intensity pixel values. This information is then employed to determine the seed type involved in the object, including additional support information such as the seed's diameter and area.

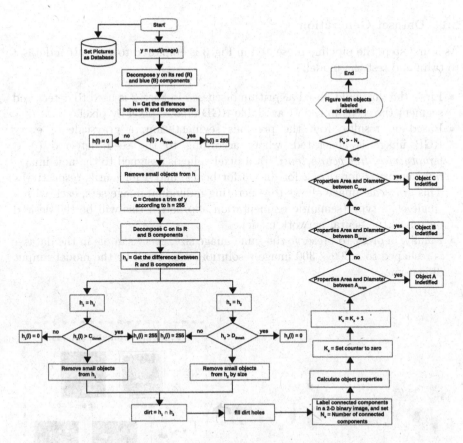

Fig. 2. Threshold-based segmentation process block diagram.

3   Detection Using Neural Networks

As mentioned previously, the threshold-based segmentation algorithm allows recognizing the pixels related to the proposed set of seeds. Nevertheless, in practice, the algorithm can suffer if some type of noise is presented in the images (such as changes to illumination, occlusions between seeds, etc.), requiring the modification of several threshold values. As neural networks have the capacity to generalize tasks, an alternative model is developed based on a semantic segmentation scheme. In this task type, the model must learn to classify every pixel p in the input image T according to a set of n defined classes $c \in [0, 1, ..., n]$. For better practical generalization, two additional classes are considered: the first employed to recognize empty space and the second to recognize dirt (any object which is neither empty space nor a seed). Thus, the full set of classes $c \in [empty, rice, dirt, bean, lentil]$ is considered for the semantic segmentation task.

3.1 Dataset Generation

As a first step, the pipeline presented in Fig. 3 is employed to generate a dataset
to train and test the model:

- First, the threshold-based algorithm discussed in Sect. 2 is used to detect and
  segment the seeds in the 41 available RGB images pixel by pixel.
- Based on results from the previous segmentation, a grayscale of every
  RGB image is generated, where according to the seed detected ($c \in$
  $[empty, rice, dirt, beans, lentils]$), a pixel value is assigned to the new image;
  0 for empty, 1 for rice, 2 for dirt, 3 for beans, and 4 for lentils respectively.
  This grayscale image allows to generating individual image sets, each with a
  single class type (semantic segmentation approach); this will be the desired
  output of the neural network model.
- Finally, in order to preserve the same image size, every sample in the dataset
  is reshaped to a 300×300 image resolution, which will be the model's input
  dimension.

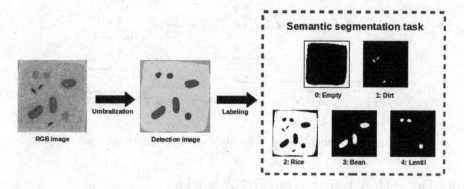

Fig. 3. Dataset building process. The threshold-based algorithm classifies the pixels
in every input image according to its seed type, generating a labeled grayscale image,
which is used to train a neural network model according to a semantic segmentation
scheme.

Once the images are processed and labeled, a data augmentation process
is carried out to augment the number of samples and allow the model to be
robust against different noise types by employing techniques such as mirroring,
cropping, and synthetic noise addition. At the end of the process, 410 labeled
RGB images are generated, distributed into 369 for training and 41 for testing.

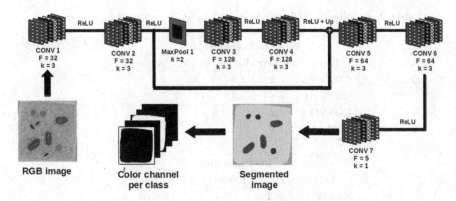

Fig. 4. Semantic segmentation neural network structure. The network allows classification of every pixel in the input images according to its seed type; empty (0), dirt (1), rice (2), beans (3), and lentils (4).

3.2 Model Definition

Once the dataset is generated considering a semantic segmentation task, a model should be defined in the required pipeline; the model will receive an RGB image as input and will generate a set of grayscale images according to the number of seeds or classes to be detected, i.e., every image will include pixels between the range $p\ in[0,1]$ according to the presence (or absence) of every defined class.

In order to find the best model in terms of accuracy and inference speed, several architectures based on ResNet blocks [8] are tested by varying the number of convolutional layers and filters. Every model was implemented in Python-Keras [13] using the image-segmentation package [5] and trained for ten epochs using Adam [14] as the learning law with a categorical cross-entropy loss function. A summary of the accuracy of every model in the validation dataset is presented in Table 1, while the architecture of each is presented in Table 2. As can be seen, the fifth model achieves the best performance, reaching an overall accuracy of 92% in the validation dataset.

Table 1. Average performance per class on validation dataset using different model architectures.

Performance per class						
Model	Params	Empty	Rice	Dirt	Beans	Lentil	Full
M1	1367	0.00	**1.00**	0.00	0.00	0.00	0.20
M2	5249	0.95	0.99	0.40	0.96	0.96	0.85
M3	20573	0.94	0.99	0.48	0.97	0.97	0.87
M4	81461	0.93	0.99	0.58	**0.98**	0.98	0.89
M5	324197	**0.98**	0.99	**0.66**	**0.98**	**0.99**	**0.92**
M6	471685	0.95	0.99	0.64	0.97	0.98	0.91

Table 2. Architecture of neural network models employed for training.

M1 - Model architecture				
Layer	Filters	Kernel	Stride	Padding
CONV 1-2	2	3	1	1
ResNet 1	8	3	1	1
CONV 7-8	4	3	1	1
CONV 9	6	1	1	0

M2 - Model architecture				
Layer	Filters	Kernel	Stride	Padding
CONV 1-2	4	3	1	1
ResNet 1	16	3	1	1
CONV 7-8	8	3	1	1
CONV 9	6	1	1	0

M3 - Model architecture				
Layer	Filters	Kernel	Stride	Padding
CONV 1-2	8	3	1	1
ResNet 1	32	3	1	1
CONV 7-8	16	3	1	1
CONV 9	6	1	1	0

M4 - Model architecture				
Layer	Filters	Kernel	Stride	Padding
CONV 1-2	16	3	1	1
ResNet 1	64	3	1	1
CONV 7-8	32	3	1	1
CONV 9	6	1	1	0

M5 - Model architecture				
Layer	Filters	Kernel	Stride	Padding
CONV 1-2	32	3	1	1
ResNet 1	128	3	1	1
CONV 7-8	64	3	1	1
CONV 9	6	1	1	0

M6 - Model architecture				
Layer	Filters	Kernel	Stride	Padding
CONV 1-2	32	3	1	1
CONV 3-4	64	3	1	1
ResNet 1	128	3	1	1
ResNet 2	64	3	1	1
CONV 9-10	32	3	1	1
CONV 11	6	1	1	0

This model comprises seven convolutional layers, including a single ResNet block in the encoder (Fig. 4) and an output layer with five filters according the required set of classes. Once trained, the neural network can directly classify every pixel in the RGB images according to its seed type; empty (no seed), dirt, rice, beans, and lentils, achieving an inference speed of 70 ms using an Nvidia GeForce GTX 1070. As can be seen in Fig. 5, the network is able to efficiently classify the seeds even when there is an occlusion between different objects, proving its robustness.

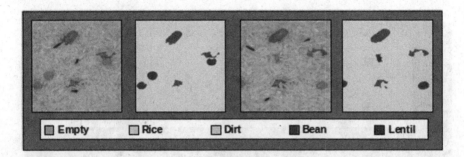

Fig. 5. Neural network segmentation output for some validation images where occlusions are present between distinct types of seeds. As can be seen, the model is able to efficiently detect and classify the pixels related to the different seeds involved in the scenes.

A confusion matrix of the model is also presented in Table 3 where most of the classes achieve a performance above 90% except for the dirt class, which gets the lowest value (around 66.14%). This effect may be due to this class having the lowest number of pixels compared with the other classes despite the data augmentation process (approximately 1.1% of the pixels in the dataset). In practice this value can be improved by augmenting the dataset with images containing a higher proportion of dirt against other seed types.

Table 3. Confusion matrix produced by the trained model using the predicted pixels from the validation dataset.

Inference Confusion Matrix						
Dataset		Percentage of class detections				
Class	N. Pixels	Empty	Rice	Dirt	Beans	Lentil
Empty	150465	98.01	1.99	0.00	0.00	0.00
Rice	3108739	0.33	99.05	0.38	0.10	0.14
Dirt	42520	0.00	6.61	66.14	23.03	4.22
Beans	227682	0.05	0.00	0.62	97.72	1.61
Lentil	70594	0.00	0.00	0.06	0.34	99.60

As a comparison for the pipelines in Fig. 6, the neural network and the threshold-based algorithm are employed to classify the seeds in four new samples; the first row presents some samples for dataset images used to verify the proposed methodology. In the second row, the images are processed using the threshold-based algorithm. Finally, the third column displays results for the proposed semantic segmentation neural network model, where the classification task for the distinct types of classes (empty (no seed), rice, dirt, beans, and lentils) are successfully achieved. Based on these results, the methodology proposed in Sect. 2 and Sect. 3 might be repeatable in other applications.

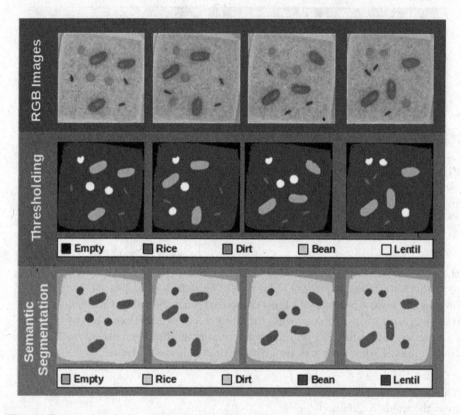

Fig. 6. Experimental algorithm results. Top: Validation input RGB images, Middle: Classification employing the threshold algorithm, Bottom: Classification employing a convolutional neural network with a semantic segmentation scheme.

4   User Interface Development

In this section, an application called *SeedApp* is developed to enable the automatic detection of the seeds. This interface is programmed in *Python (v3.9.7)* using the *KivyMD (v1.1.1)* and *Kivy (v2.1.0)* libraries.

4.1   SeedApp Description

As can be seen in Fig. 7, the *SeedApp* has a minimalist interface with just two buttons: The *load image* button allows loading images from disk, while *process image* starts the processing pipeline. Additionally, the home environment has two icons in the superior right and left corners, *Menu* and *Change Colors*, respectively.

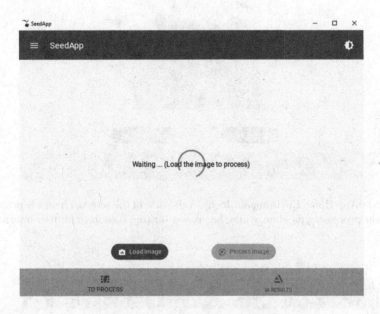

Fig. 7. SeedApp Home Environment. The app has a minimalist interface with a set of buttons to facilitate user interaction, including buttons for uploading and processing images, as well as for customizing the interface.

Figure 8 shows the color's interface changed using the icon *Change Colors*. When the user presses the upload button, a search explorer window pops up to facilitate selection of the image processing path, as shown in Fig. 9. Once an image is selected, a preview of same is displayed and the *process* button is enabled for interaction.

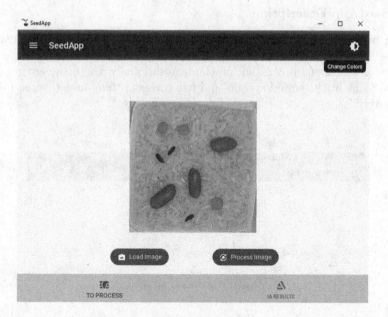

Fig. 8. SeedApp Home Environment Icons. A preview of the selected image is presented prior to the processing pipeline, while the *process* buttom is enabled for user interaction.

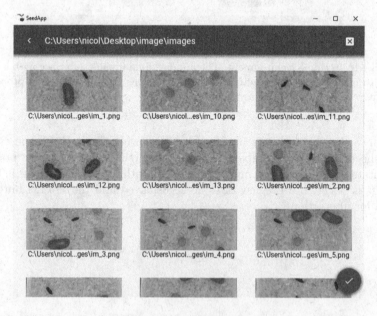

Fig. 9. SeedApp Pop Up Search Window. A friendly interface is provided to help the user to navigate between the images to be segmented.

Once the user presses the button, the tool initializes seed segmentation and classification. Once processed, the app presents additional information, such as the type and number of seeds found, including different textual and visual feedback for the user (Fig. 10).

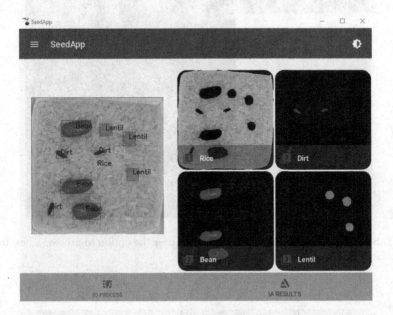

Fig. 10. SeedApp IA Results. The app provides the user the result of the segmentation using different visual interfaces: First, every seed is labeled with a bounding box, including its label (Left). Second, every seed type found is labeled and isolated in individual images. In practice, this allows the user to obtain a richer set of features from the segmentation result.

As an additional feature, the SeedApp has an icon to access the menu (Fig. 11); this environment allows the user to change between the options *home* and *recent*.

- The *home* button allows bringing up the image loading screen, where the user can process as many images as are required.
- The *recent* option (illustrated in Fig. 12) shows the user all results obtained since the app was initialized. This affords the user a formal record of the segmentation, classification, and seed count obtained by the application across different image samples.

For additional multimedia resources obtained with the SeedApp, as well as to see its behavior in real-time, please consult [31].

Fig. 11. SeedApp Menu Browser. This gives the user the option to process a new image or to consult the detection history.

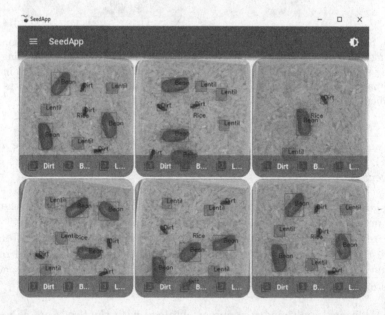

Fig. 12. SeedApp Recent Option. This feature allows the user to visualize the detection history, including a count of the seeds found in each input image since the app was launched.

5   Conclusions

In this work, a set of vision-based algorithms are proposed to detect and classify a set of seeds in mixed containers. As a first approach, an algorithm based on pixel intensity thresholds was developed to efficiently detect three types of seeds: rice, beans, and lentils. Experimental results illustrate that the proposed algorithm efficiently executes the proposed detection tasks.

As a second approach, a convolutional neural network was developed using a semantic segmentation scheme. By employing the samples provided by the threshold algorithm as a training dataset, the trained model can detect and classify a total set of five classes (including empty space, and dirt) from RGB images. Additionally, a user interface was developed in an application to enable users to run the full pipeline on new image samples and obtain information related to the number and type of seeds found.

As future work, the authors will extend the proposed algorithms to consider other seed types, images in more complex environments, and different object distributions.

References

1. Ben-Daya, M., Hassini, E., Bahroun, Z., Banimfreg, B.: The role of internet of things in food supply chain quality management: a review. Qual. Manag. J. **28**(1), 17–40 (2020)
2. Beyerer, J., Puente León, F., Frese, C.: Machine Vision: Automated Visual Inspection: Theory, Practice and Applications. Springer, Heidelberg (2015)
3. Constante, P., Gordon, A., Chang, O., Pruna, E., Acuna, F., Escobar, I.: Artificial vision techniques to optimize strawberry's industrial classification. IEEE Lat. Am. Trans. **14**(6), 2576–2581 (2016)
4. Dana, W., Ivo, W.: Computer image analysis of seed shape and seed color for flax cultivar description. Comput. Electron. Agric. **61**(2), 126–135 (2008)
5. Divamgupta: Image Segmentation Keras: Implementation of Segnet, FCN, UNet, PSPNet and other models in Keras. Github (2020). https://github.com/divamgupta/image-segmentation-keras. Accessed 26 Dec 2020
6. Galdelli, A., et al.: A novel remote visual inspection system for bridge predictive maintenance. Remote Sens. **14**(9), 2248 (2022)
7. Golnabi, H., Asadpour, A.: Design and application of industrial machine vision systems. Robot. Comput.-Integr. Manuf. **23**(6), 630–637 (2007)
8. He, K., Zhang, X., et al.: Deep residual learning for image recognition. In: Proceedings of the IEEE Conference on Computer Vision and Pattern Recognition, pp. 770–778 (2016)
9. Ismail, N., Malik, O.: Real-time visual inspection system for grading fruits using computer vision and deep learning techniques. Inf. Process. Agric. **9**(1), 24–37 (2022)
10. Jaffery, Z., Dubey, A.: Scope and prospects of non-invasive visual inspection systems for industrial applications. Indian J. Sci. Technol. **9**(4), 1–11 (2016)
11. Jampílek, J., Král'ová, K.: Application of nanotechnology in agriculture and food industry, its prospects and risks. Ecol. Chem. Eng. S **22**(3), 321–361 (2015)

12. Jhawar, J.: Orange sorting by applying pattern recognition on colour image. Procedia Comput. Sci. **78**, 691–697 (2016)
13. Keras-team: Keras: Deep Learning for Python. Github (2020). https://github.com/keras-team/keras. Accessed 26 Dec 2020
14. Kingma, D., Ba, J.: Adam: a method for stochastic optimization. arXiv:1412.6980 (2014)
15. Kumar, S., Patil, R., Kumawat, V., Rai, Y., Krishnan, N., Singh, S.: A bibliometric analysis of plant disease classification with artificial intelligence using convolutional neural network. Libr. Philos. Pract. **5777**, 1–14 (2021)
16. LeCun, Y., Bengio, Y., Hinton, G.: Deep learning. Nature **521**(7553), 436–444 (2015)
17. Lu, Y., Yi, S., Zeng, N., Liu, Y., Zhang, Y.: Identification of rice diseases using deep convolutional neural networks. Neurocomputing **267**, 378–384 (2017)
18. Macias-Garcia, E., Galeana-Perez, D., et al.: Multi-stage deep learning perception system for mobile robots. Integr. Comput.-Aided Eng. (Preprint) 1–15 (2020)
19. Mansuri, L., Patel, D.: Artificial intelligence-based automatic visual inspection system for built heritage. Smart Sustain. Built Environ. **11**, 622–646 (2021)
20. Manzoor, M., et al.: A narrative review of recent advances in rapid assessment of anthocyanins in agricultural and food products. Front. Nutr. **9** (2022)
21. Misra, N., Dixit, Y., Al-Mallahi, A., Bhullar, M., Upadhyay, R., Martynenko, A.: IoT, big data and artificial intelligence in agriculture and food industry. IEEE Internet of Things (2020)
22. Negrete, J.: Artificial vision in Mexican agriculture, a new technology for increase food security. Manag. Econ. J. 381–398 (2018)
23. Newman, T., Jain, A.: A survey of automated visual inspection. Comput. Vis. Image Underst. **61**(2), 231–262 (1995)
24. Ravikumar, S., Ramachandran, K., Sugumaran, V.: Machine learning approach for automated visual inspection of machine components. Expert Syst. Appl. **38**(4), 3260–3266 (2011)
25. Redmon, J., Farhadi, A.: YOLOv3: an incremental improvement. arXiv:1804.02767 (2018)
26. Romualdo, L., et al.: Use of artificial vision techniques for diagnostic of nitrogen nutritional status in maize plants. Comput. Electron. Agric. **104**, 63–70 (2014)
27. Santos-Gomes, J., Rodrigues Leta, F.: Applications of computer vision techniques in the agriculture and food industry: a review. Eur. Food Res. Technol. **235**(6), 989–1000 (2012)
28. Sarkar, N.R.: Machine vision for quality control in the food industry. In: Instrumental Methods for Quality Assurance in Foods, pp. 167–187. Routledge (2017)
29. Schmidhuber, J.: Deep learning in neural networks: an overview. Neural Netw. **61**(1), 85–117 (2015)
30. Suarez, O., Hernández Díaz, N., Pardo García, A.: A real-time pattern recognition module via Matlab-Arduino interface. In: 18th LACCEI International Multi-Conference for Engineering, Education, and Technology, pp. 1–8 (2020)
31. Suarez, O., Macias-Garcia, E., Vega, C., Carrillo, Y., Hernandez, N., Garrido, V.: Design of a segmentation and classification system for seeds detection based on pixel intensity thresholds and convolutional neural networks: supplementary material: Seedapp v1.0.0 an app using kivy and kivymd framework. https://youtu.be/1li2PjH_Z9A
32. Szeliski, R.: Computer Vision: Algorithms and Applications. Springer, Heidelberg (2010)

33. Voulodimos, A., Doulamis, N., Doulamis, A., Protopapadakis, E.: Deep learning for computer vision: a brief review. Comput. Intell. Neurosci. **2018** (2018)
34. Yu, C., Wang, J., et al.: BiSeNet: bilateral segmentation network for real-time semantic segmentation. In: Proceedings of the European Conference on Computer Vision, pp. 325–341 (2018)

Classification of Focused Perturbations Using Time-Variant Functional Connectivity with rs-fmri

Catalina Bustamante[1,2]([✉]) [iD], Gabriel Castrillón[2] [iD],
and Julián Arias-Londoño[3] [iD]

[1] Universidad de Antioquia, Medellín, Colombia
catalina.bustamante@udea.edu.co
[2] G. de Investigación en Imágenes Médicas SURA, Medellín, Colombia
jgcastrillong@sura.com.co
[3] Universidad Politécnica de Madrid, Madrid, Spain
julian.arias@upm.es

Abstract. Focal disturbances in the cerebral activity modulated by transcranial magnetic stimulation (TMS) produce alterations in brain connectivity at the global level. Those effects can be studied using time-varying functional connectivity (TVFC) based on functional magnetic resonance imaging at rest (rs-fMRI). The characteristics of these alterations could be modeled using machine learning algorithms for patient classification. This study used hidden Markov models (HMM) to evaluate temporal variations in functional connectivity after stimulation of two different brain areas (Frontal and Occipital). We modeled the dynamics of 15 resting-state networks in 12 states by calculating the fractional occupancy, mean lifetime, and interval time of each state. We then compared the difference between fMRI sessions, PRE, and POST-stimulus, observing significant differences for both conditions, especially after frontal stimulation. Finally, generative models based on HMM were trained, to classify PRE-stimulus and Occipital stimulus with an accuracy of 83%, PRE-stimulus and Frontal stimulus with an accuracy of 85%, and Occipital and Frontal stimulus with 65% accuracy. This finding could be extended to the characterization of pathologies where local disturbances have a global impact on functional connectivity, such as Epilepsy.

Keywords: rs-fMRI · TMS · HMM · TVFC

1 Introduction

Transcranial magnetic stimulation (TMS) is a non-invasive brain stimulation technique that disturbs the neuronal activity in cortical areas due to electrical

J. Arias-Londoño—started this work at the Antioquia University and finished it supported by a María Zambrano grant from the Universidad Politécnica de Madrid, Spain.

currents induced externally by a variable magnetic field. The local disruption of the neuronal activity or virtual lesion [17] allows the study of brain functional networks, a group of disaggregated synchronized regions, by modulating their interregional synchronization [5]. Focal perturbations in the brain activity generate alterations in the global functional connectivity [2], which can be measured using resting-state functional magnetic resonance imaging (rs-fMRI), and being analyzed dynamically over time using time series analysis techniques such as time-varying functional connectivity (TVFC). Behavior and cognitive changes induced experimentally through TMS may provide causal evidence for the relevance of TVFC [10]. Using TMS, it is possible to alter the patterns of cognitive activity selectively, and functional connectivity in related brain networks, as well as the interaction between them [5]. Therefore, in conjunction with rs-fMRI, TMS offers an opportunity to perturb neuronal information processing and measure its effects. Additionally, it offers the possibility to characterize their effects within the same subject, minimizing the between-subject variability [17] and identifying the effect of focal disturbances in the brain functional connectivity that could yield to optimized models for detecting diseases such as epilepsy. Concerning epilepsy, the correct diagnosis of it is essential in the process of treatment, and pre-surgical evaluation. in this sense, providing a better understanding of the brain networks altered by a particular focus could improve the diagnosis, and the identification of possible biomarkers and targets for the treatment of those patients [1].

The most common methods to estimate TVFC are based on sliding windows, in which functional connectivity is calculated within time windows throughout the entire time series [10]; however, this approach imposes assumptions about the optimal size of the window [10]. In contrast, generative methods, such as hidden Markov models (HMM), can evaluate temporal variations in functional connectivity, without defining an analysis window and using all the data for model estimation. By using HMM for TVFC, brain activity is modeled as a discrete set of states, characterized by different connectivity patterns [19]. It allows the characterization of brain activity over time, assessing the hierarchical organization of brain states and recurrent patterns of brain functional connectivity, as well as evaluating changes in this behavior under different conditions and pathologies. Additionally, HMM offers a flexible way to track stationary and abrupt changes in brain connectivity [22].

HMMs are probabilistic models in which the observed data are assumed to be generated by a process moving through unobservable states [10]. Fitting an HMM involves evaluating the probability of a sequence of observations given a model, determining the best sequence of states of the model that can explain the observations, as well as estimating and fitting the parameters of the model [14]. For fMRI, HMM states might represent a brain network with a unique pattern of functional connectivity [20]; where the transition probabilities represent how likely a transition between different brain states is, and the estimation of the active state represents which state is manifested in each unit of time [10]. These parameters can provide an overview of the temporal dynamics between brain areas and brain states. Therefore, the selection and comparison of different model parameters, allows the evaluation of which model best describes the data and, consequently, having hypotheses about its generation process [10].

Here, we evaluated the use of HMM over a dataset of rs-TMS-fMRI. We compared the differences between state sequences and found some differences between different areas of stimulation. This finding could be extended to the evaluation of the effects of global functional connectivity over time, in local perturbation scenarios.

2   Materials and Methods

2.1   Data Acquisition

We used a freely available dataset[1] [2] that contains rs-TMS-fMRI and anatomical images of 23 healthy participants (14 females, 25.56 years mean age). Images were acquired during different days in which a low-frequency (1HZ) rTMS was applied on three different cortical regions: prefrontal (FRO), occipital (OCC), and temporoparietal control (CTR); rs-fMRI images were acquired before (PRE) and after stimulation (POST). To compare the effect of stimulation in two different functional regions, we selected a somatomotor area, the OCC session, and a higher-cognitive area of interest, the FRO session, according to [2].

Functional MRI images were acquired on a 3T Philips Ingenia MRI scanner (40 slices, TR = 1250 ms, TE = 30 ms, FA= 70°, FOV = 192 mm × 192 mm, matriz size = 64 × 64, voxel size = 3 mm × 3 mm × 3 mm), 600 volumes were acquired with each session lasting 12.35 min, during the session, participants were asked to lay relaxed, awake, and with open eyes. They acquired in addition an anatomical image for each session (170 slices; repetition time, TR = 9 ms; echo time, TE = 3.98 ms; flip angle, FA = 8°; field of view, FOV = 256 mm × 256 mm; matrix size = 256 × 256; voxel size = 1 mm × 1 mm × 1 mm).

In order to minimize variability and obtain more conclusive and specific results associated with combined TMS-fMRI studies, the dataset's publication followed the methodological considerations proposed previously [4, 13]: (i) electric-field neuronavigation to identify functional target regions in each individual [12, 15], (ii) neuroimaging to evaluate the effects of stimulation beyond the stimulated network, and (iii) stimulation of a control region (in contrast to sham stimulation) to test the functional specificity of the target areas.

2.2   Image Processing

Functional and anatomical images were processed using the Configurable Pipeline for the Analysis of Connectomes[2] The anatomical images were brain extracted using AFNI-3dSkullStrip; segmented automatically into white matter, gray matter, and CSF masks using FSL-FAST, and normalized to the standard

---

[1] https://openneuro.org/datasets/ds001927/versions/2.0.2.
[2] C-PAC https://fcp-indi.github.io/.

MNI152 2 mm space. The functional images were registered to the anatomical images, slice-timing corrected, realigned, motion-corrected to a reference mean image using AFNI-3dvolreg, and skull-stripped using AFNI-3dAutomask. Later, the nuisance signals were regressed out, and the signal was bandpass-filtered (0.01 to 0.1 Hz). Subsequently, spatial smoothing was performed using a Gaussian filter with FWHM of 4 mm, and the time series was extracted by averaging the timeseries of every region of interest (ROI) from a parcellation of 400 regions defined by Shaeffer [16]. In this parcellation, each ROI is associated with one of 17 different functional networks as shown in (Fig. 1) which are defined as subsets of synchronized brain regions defined previously by Yeo [21]. The ROIs located in the Limbic network were removed due to high noisily signals on those regions in fMRI images, resulting in 376 ROIs with 15 networks association.

2.3 Hidden Markov Model

To implement the Hidden Markov Models, we assumed that the time series could be described as a sequence of a finite number of independent not observable states [10]. The models were implemented using the hmmlearn Python library, version 0.27[3] First, we defined the pre-stimulus image as the average of the two PRE session images. Then, we created an average signal per network for each session (15 variables with 600-time points); signals were afterward z-scored. A GaussianHMM model was trained using only the average signal of Pre-stimulus sessions; time series were concatenated time-wise across all subjects generating an input matrix of size (23 * 600) x 15; a full rank covariance matrix was used, and 12 states were selected according to the selection used in previous works [9,20]. We calculated the optimal number of states for our data using the Bayesian Information Criterion (BIC) and the Akaike Information Criterion (AIC), getting a number of 9 and 21 states respectively as the optimal number; However, we set this parameter to 12, following previous studies which have used HMM in bigger datasets [20].

Hidden State Sequences. The hidden most probable state sequence for each session was calculated using the Viterbi algorithm. For every session of each subject, (1 averaged session of Pre-stimulus, 1 session for Post-FRO, 1 session for Post-OCC) the Viberbi paths were calculated, showing the most probable state in each of the 600 time points. Based on these states, three characteristics of the sequences for each state were then calculated: Fractional occupancy (FO) which corresponds to the proportion of time spent in each state; mean lifetime (MLT) as the average time of the visits to each state and the mean interval time (IT), which represents the average time between visits for each state; these steps are illustrated in figure Fig. 1.

HMM Based Generative Model. From the time series data, a generative classification model was built. Unlike discriminative models in which prediction

---

[3] Available at https://pypi.org/project/hmmlearn.

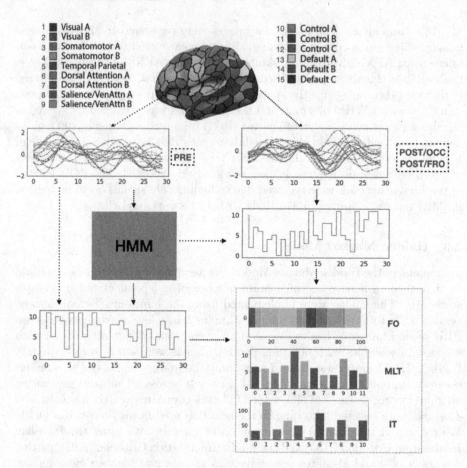

Fig. 1. Summary diagram of the procedure. The average signals from Yeo parcellation that groups the ROIS in different networks are obtained. With the data of the Pre-stimulus session, an HMM model is trained. From this model, the most probable state sequences for each subject in the PRE and POST sessions are extracted. From these sequences, three variables were calculated and compared between groups.

is made based on conditional probability, generative models focus on the distribution of a data set to return the probability of a sample, explaining how the data was generated [6]. Generative models have shown better performance for small sample sizes compared to discriminative models [11]. Independent HMM models were trained for each class; from these models, the probability of a test sample being generated by each of the models was evaluated, and the class with the highest likelihood was selected.

3   Experiments

3.1   Statistical Analysis

We performed two-sided Wilcoxon signed-rank tests for each of the variables to compare the differences in characteristics of the state sequences between Pre-stimulation and Post-stimulation sessions and corrected for multiple comparisons using FDR Benjamini & Hochberg correction.

3.2   Classification Model

Three generative models were trained to classify PRE vs OCC, PRE vs FRO, and OCC vs FRO. Given the high dimensionality of the data characteristics (376) and the few samples available (23 in each class), a dimension reduction is necessary before training the model. For this purpose, principal component analysis (PCA) was used.

Selection of the model parameters was carried out with a search through a grid of values (GridSearch) in which the following options were included: Number of states: [8–17], Covariance matrix type: [full, diagonal], HMM Type: [Gaussian, Mixture of Gaussians], Number of Gaussians for the mixture model: [2–10], Variance explained by PCA: [0.70–1.0]

The model performance estimation process was carried out using cross-validation; Since the PRE and POST groups correspond to the same subject, cross-validation was performed, leaving a subject out from both classes (LeveOneGroupOut). Taking into account that each class of the model has the same number of subjects, the set of performance metrics used for evaluation purposes include accuracy, confusion matrices, ROC curves, and the area under the ROC curve (AUC).

4   Results

4.1   Statistical Analysis

The statistical analysis, comparing the Pre-stimulus group with the Post-stimulus-FRO and Post-stimulus-OCC group on the three variables, is shown in Table 1. Figure 2b shows the comparison between the three groups, only for the states which had a significant difference in any of both post-stimulus groups. For the FO variable, in the OCC stimulation, no significant differences were found after correcting for multiple comparisons in any of the 12 states ($p_{FWD} > 0.05$). However, for FRO stimulation, differences were found in six states ($p_{FWD} \leq 0.05$), which were named according to their activation pattern: "Base-high", in which all networks have a similar activation with a tendency to high; "Low" in which all networks show a pattern of low activation; "Base" in which all networks showed the baseline activation and "Complex-functions" "A, B and C", with variable activation in networks related to complex cognitive functions. The last three states seem to be related, showing antagonistic activation in all networks, except for the visual network, which remains with low activation in all three states. Furthermore, the fraction occupancy was lower for both stimulation sessions in the states in which all the networks had a similar activation (high, low, or average), but it was only statistically significant for the FRO stimulation. In contrast, states named "complexity-functions" showed an increase in the percentage of time they were active; However, no significant differences were found in the mean lifetime, indicating a change in the number of visits to these states, but not in the duration of the visits. On the other hand, for the Mean Lifetime variable, the value was significantly inferior for the state named "Base-high" for both groups after correcting for multiple comparisons ($p_{FWD} \leq 0.05$).

Finally, for the IT variable, significant differences were found only for the Post-FRO group in the state named "High", in which all states present a high activation pattern; the augmented interval time is consistent with the reduction in FO and MLT, showing a state less visited for session after stimulation in FRO region mainly.

Table 1. Results of the statistical tests

State	OCC			FRO		
	FO	*MLT*	*IT*	*FO*	*MLT*	*IT*
Base-high	0.43	0.02	0.58	0.03	0.01	0.46
Complex-funct A	0.30	0.72	0.22	0.054	0.68	0.43
Base-low	0.89	0.28	0.79	0.23	0.057	0.43
Low-BaseDMN	0.13	0.56	0.42	0.40	0.65	0.55
Complex-funct B	0.30	0.56	0.42	0.01	0.21	0.28
Complex-funct C	0.30	0.72	0.57	0.054	0.68	0.096
Base	0.89	0.72	0.96	0.054	0.21	0.096
Low	0.66	0.28	0.61	0.03	0.057	0.096
DMN	0.82	0.11	0.58	0.51	0.07	0.55
Visual-low	0.29	0.16	0.58	0.18	0.21	0.43
High	0.29	0.11	0.22	0.24	0.21	0.03
Visual-high	0.44	0.11	0.58	0.24	0.65	0.14

[a] $p_{FWD} \leq 0.05$ are highlighted in blue

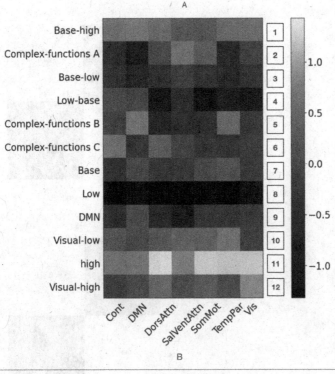

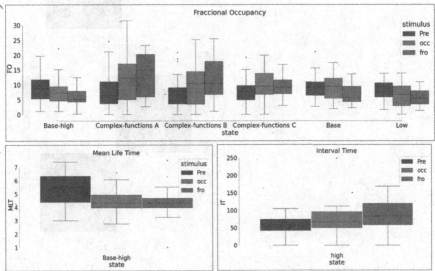

Fig. 2. A. Mean intensity for each network in each state; the 15 networks were grouped into 7 to simplify the analysis. B. Comparison of the values of fractional occupancy, mean lifetime, and interval time. Only the states that were significantly different between the Pre sessions and any of the Post sessions are shown.

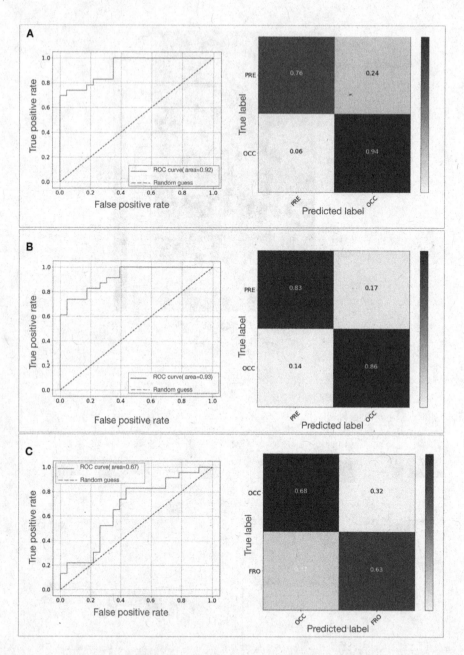

Fig. 3. A. Mean intensity for each network in each state; the 15 networks were grouped into 7 to simplify the analysis. B. Comparison of the values of fractional occupancy, mean lifetime, and interval time. Only the states that were significantly different between the Pre sessions and any of the Post sessions are shown.

4.2 Classification Analysis

The highest accuracy value for the classification of PRE-stimulus versus POST-OCC stimulus was 0.83 95% CI (confidence interval) [0.70–0.91]; corresponding to an HMM with 3 Gaussian mixtures, 9 states, and a percentage variance explained with PCA of 90%. (Fig. 3) A, shows the ROC curve and the confusion matrix for this model where an AUC value of 92% and correct classification of 94% for OCC and 76% for PRE are observed.

For the classification between PRE-stimulus and POST-FRO stimulation, the best model was found with an overall accuracy value of 0.85 95% CI (confidence interval) [0.72–0.93]; for an HMM with 4 Gaussian mixtures, 9 states, and 70% of explained variance with PCA. AUC value was 93% with an accuracy of 83% for PRE-stimulus and 86% for the Post-FRO group (Fig. 3) B.

Finally, for the classification between OCC and FRO stimulation, the highest accuracy value was obtained for 65% 95% CI [51%–78%]; with an HMM with a full covariance matrix, 10 states, and without PCA reduction. Figure 3) C, shows the ROC curve and the confusion matrix for this model, the area under the curve was 67%, and accuracies of 68% for OCC-stimulus and 63% for FRO-stimulus group.

5   Discussion and Conclusions

Localized disturbances in brain activity, caused by external stimulation using TMS, generate alterations that extend to different areas of the brain, outside the stimulated area [2]. These changes also vary over time since brain activity is dynamic. The methods used to study the brain with controlled stimulation, as in the case of TMS, could later be used to evaluate pathologies such as epilepsy, in which alterations in brain activity also extend to other areas outside the epileptogenic focus [3].

We evaluated the temporal dynamics of fMRI activity after TMS stimulation using HMM models and found that stimulation generates changes in the pattern of visits to the different states; these changes were mainly seen on FRO stimulation. Comparing the results obtained with the same dataset using static connectivity analysis in the previous work of Castrillon [2], we also did not find alterations at the local level of the visual area for OCC stimulation, since changes in visits to states with activation of this network were not significantly different. The effect of FRO stimulation was more widespread than OCC stimulation in Castrillon's work; this fact could explain why we observed more changes in the characteristics of the sequence of states after FRO stimulation.

A different behavior was observed for the states that have a similar activation in all the networks; compared to states characterized by activations of networks involved in more complex brain functions. These changes showed a decrease in the proportion of time that the Post group visits the states with equal network activation; however, there is no significant change in the time interval between visits or in the time spent in those states (only in one of them). Therefore, in general, the number of times these states are visited has decreased, but not necessarily the duration of these visits. The time in which the equal network activation states are not visited seems to be replaced by states in which the activation of more complex networks is prioritized. These networks additionally seem to be complementary, showing an antagonistic activation pattern of some networks.

At the subject level, from generative models such as HMM, it was possible to detect focused disturbances generated by stimulation with TMS; the pure BOLD signals of the 376 ROIS were used; carrying out the analysis with dimension reduction through PCA; this classification had its best performance for stimulation in the frontal area (85%) with similar results (83%) for the occipital area. The classification between the two stimulated areas had lower results (65%); however, it is within the same range (67%) of results previously found by the authors of a previous study where the same data was used for a static connectivity analysis [2]. The model evaluated achieved a moderate classification accuracy inherent to neuroimaging datasets, which are usually underpowered, having small effect sizes due to the small sample size. However, when studying clinical populations or focused interventions, such as brain stimulation, small-scale datasets are often unavoidable [7]. Furthermore, the prediction accuracy reported in the literature often decreases as sample sizes increase due to the large variance of neuroimaging datasets [18].

Several limitations must be taken into account. The sample size is small, leading to high variability and high confidence intervals in validation through cross-validation [18]. In the HMM approach used, intensity-based states are generated; There are other approaches based directly on functional connectivity values over time [8]; which could present states with more distinguishable patterns. Additionally, during the functional connectivity analysis, only one model was trained with the PRE sections, so it is assumed that the POST sessions share the same set of states, changing the duration and frequency of the visits; however, stimulation can generate states with different activation patterns and therefore a separate model could be trained for these sessions, although it would complicate the analyzes by not being able to make direct comparisons between the activation patterns of the states. These experiments could be the target of further analysis.

We used HMM to explore the dynamic changes that occur in a brain after TMS stimulation and found some differences mainly for FRO stimulation. Additional studies with more subjects and other types of datasets should be done to evaluate the technique with other types of localized disturbances such as epilepsy.

References

1. Abela, E., Rummel, C., Hauf, M., Weisstanner, C., Schindler, K., Wiest, R.: Neuroimaging of epilepsy: lesions, networks, oscillations. Clin. Neuroradiol. **24**(1), 5–15 (2014). https://doi.org/10.1007/s00062-014-0284-8

2. Castrillon, G., Sollmann, N., Kurcyus, K., Razi, A., Krieg, S.M., Riedl, V.: The physiological effects of noninvasive brain stimulation fundamentally differ across the human cortex. Sci. Adv. **6** (2020). https://doi.org/10.1126/sciadv.aay2739

3. Elshoff, L., et al.: Dynamic imaging of coherent sources reveals different network connectivity underlying the generation and perpetuation of epileptic seizures. PLoS ONE **8**, 1–11 (2013). https://doi.org/10.1371/journal.pone.0078422

4. Fox, M.D., Halko, M.A., Eldaief, M.C., Pascual-Leone, A.: Measuring and manipulating brain connectivity with resting state functional connectivity magnetic resonance imaging (fcMRI) and transcranial magnetic stimulation (TMS). NeuroImage **62**, 2232–2243 (2012). https://doi.org/10.1016/j.neuroimage.2012.03.035, http://dx.doi.org/10.1016/j.neuroimage.2012.03.035

5. Gollo, L.L., Roberts, J.A., Cocchi, L.: Mapping how local perturbations influence systems-level brain dynamics. NeuroImage **160**, 97–112 (2017). https://doi.org/10.1016/j.neuroimage.2017.01.057, http://dx.doi.org/10.1016/j.neuroimage.2017.01.057

6. Goyal, C.: Deep understanding of discriminative and generative models (2021). https://www.analyticsvidhya.com/blog/2021/07/deep-understanding-of-discriminative-and-generative-models-in-machine-learning/

7. He, T., et al.: Meta-matching as a simple framework to translate phenotypic predictive models from big to small data. Nat. Neurosci. **25**(6), 795–804 (2022). https://doi.org/10.1038/s41593-022-01059-9, https://www.nature.com/articles/s41593-022-01059-9

8. Hussain, S., Langley, J., Seitz, A.R., Peters, M.A.K., Hu, X.P.: A novel hidden Markov approach to studying dynamic functional connectivity states in human neuroimaging. bioRxiv, p. 2022.02.02.478844 (2022). https://www.biorxiv.org/content/10.1101/2022.02.02.478844v1

9. Kottaram, A., et al.: Brain network dynamics in schizophrenia: reduced dynamism of the default mode network. Hum. Brain Mapp. **40**, 2212 (2019). https://doi.org/10.1002/HBM.24519, https://aplicacionesbiblioteca.udea.edu.co:2054/pmc/articles/PMC6917018/

10. Lurie, D.J., et al.: Questions and controversies in the study of time-varying functional connectivity in resting fMRI. Netw. Neurosci. **4**, 30–69 (2020)

11. Matsubara, T.: Bayesian deep learning: a model-based interpretable approach. Nonlinear Theory Appl. IEICE **11**, 16–35 (2020). https://doi.org/10.1587/NOLTA.11.16

12. Opitz, A., Fox, M.D., Craddock, R.C., Colcombe, S., Milham, M.P.: An integrated framework for targeting functional networks via transcranial magnetic stimulation. NeuroImage 127, 86–96 (2016). https://doi.org/10.1016/J.NEUROIMAGE.2015.11.040, https://pubmed.ncbi.nlm.nih.gov/26608241/

13. Polanía, R., Nitsche, M.A., Ruff, C.C.: Studying and modifying brain function with non-invasive brain stimulation. Nat. Neurosci. **21**, 174–187 (2018). https://doi.org/10.1038/s41593-017-0054-4, http://dx.doi.org/10.1038/s41593-017-0054-4

14. Rabiner, L.: A tutorial on hidden markov models and selected applications in speech recognition. Proc. IEEE **77**, 257–286 (1989). https://doi.org/10.1109/5.18626, http://ieeexplore.ieee.org/document/18626/

15. Sack, A.T., Kadosh, R.C., Schuhmann, T., Moerel, M., Walsh, V., Goebel, R.: Optimizing functional accuracy of TMS in cognitive studies: a comparison of methods, pp. 1–15 (2008). https://doi.org/10.1162/jocn.2009.21126

16. Schaefer, A., et al.: Local-global parcellation of the human cerebral cortex from intrinsic functional connectivity MRI. Cereb. Cortex **28**, 3095–3114 (2018). https://doi.org/10.1093/cercor/bhx179

17. Sliwinska, M.W., Vitello, S., Devlin, J.T.: Transcranial magnetic stimulation for investigating causal brain-behavioral relationships and their time course. J. Vis. Exp. JoVE (2014). https://doi.org/10.3791/51735, http://www.ncbi.nlm.nih.gov/pubmed/25079670, http://www.pubmedcentral.nih.gov/articlerender.fcgi?artid=PMC4219631
18. Varoquaux, G.: Cross-validation failure: small sample sizes lead to large error bars. Neuroimage **180**, 68–77 (2018). https://doi.org/10.1016/j.neuroimage.2017.06.061
19. Vidaurre, D.: A new model for simultaneous dimensionality reduction and time-varying functional connectivity estimation. PLoS Comput. Biol. **17**, 1–20 (2021). https://doi.org/10.1371/journal.pcbi.1008580, http://dx.doi.org/10.1371/journal.pcbi.1008580
20. Vidaurre, D., Smith, S.M., Woolrich, M.W.: Brain network dynamics are hierarchically organized in time. Proc. Natl. Acad. Sci. U.S.A. **114**, 12827–12832 (2017). https://doi.org/10.1073/pnas.1705120114
21. Yeo, B.T.T., et al.: The organization of the human cerebral cortex estimated by intrinsic functional connectivity. J. Neurophysiol. **106**, 1125–1165 (2011). https://doi.org/10.1152/jn.00338.2011
22. Zhang, G., et al.: Estimating dynamic functional brain connectivity with a sparse hidden Markov model. IEEE Trans. Med. Imaging **39**, 488–498 (2020). https://doi.org/10.1109/TMI.2019.2929959

Escherichia coli: Analysis of Features for Protein Localization Classification Employing Fusion Data

Alvaro David Orjuela-Cañón[✉] [iD], Diana C. Rodriguez[iD], and Oscar Perdomo[iD]

School of Medicine and Health Sciences, Universidad del Rosario, Bogotá D.C, Colombia
alvaro.orjuela@urosario.edu.co

Abstract. Machine learning models can be used for relevance of features in classification systems. The interest in protein analysis based on biomolecular information has rapidly grown. In this case a comparison of two sources of this information was employed to determine protein localization in *Escherichia coli* cells. Models as support vector machines, artificial neural networks and random forest were compared for the prediction of protein localization. The sources of data used to train the models were the information from targeting signal and protein sequences, for determining the localization sites of the protein. A third scenario with a fusion of both sources of data was employed. Four classes were established according to the subcellular localization of the protein: cytoplasm, periplasmatic space, outer and inner membranes. Results reached values between 77% and 92% in terms of balanced accuracy. The models with better performance were based on random forest and support vector machines. In terms of features, the first source, where targeting signal was employed, was the one with best performance associated to relevance for the classification.

Keywords: Bioinformatics · Proteomics · Proteins · Machine Learning · Localization Sites Prediction · Features Relevance Analysis

1 Introduction

Bioinformatics is a field that has been grown since late nineties due to the increase of the computational resources. Simultaneously, artificial intelligence has also taken relevance due to its application in different fields. Computational intelligence (CI) can be considered as a subfield of the artificial intelligence, where the effectiveness and increasing number of applications, provide solutions that humans execute for longer periods of time. In general, the (CI) area contributes to bioinformatics, providing methods and new possibilities for analyzing the bio-data associated to this field [1, 2].

Despite bioinformatics is a huge field, there are many specific areas, as proteomics, where the analysis is carried out with information mainly extracted from peptides, polypeptides and proteins. Currently, the most known application of CI in proteomics is associated to structure prediction, where models based on deep learning has provide support on the critical assessment of protein structure prediction through the *Alphafold*

A. D. Orjuela-Cañón et al. (Eds.): ColCACI 2022, CCIS 1746, pp. 31–43, 2023.
https://doi.org/10.1007/978-3-031-29783-0_3

tool [3, 4]. Thus, the number of approaches based on CI has incremented, searching for more and new tasks, for example, drug discovery, protein engineering and *de* novo design of proteins with novel functions. These successful results drive to continue the exploration of CI techniques in the bioinformatics field.

Within the many applications, the prediction of protein localization, in subcellular sites, is one of the most important tasks in the function analysis of these macromolecules. Here, the localization is demanded from some protein information source, and then, it is possible to associate it to a specific task or function [5, 6]. In these scenarios, it is different the role of a membrane protein compared to the transportation proteins or enzymes, where its properties have changed. In addition, each time is more popular to analyze information from the protein sequence [4], where a simple primary structure with amino acid information can provide a context of the localization, function or structure. All this could not be possible without data, which is produced by robust and high-cost laboratories, where the protein characterization is obtained through processes as nuclear magnetic resonance (NMR) spectroscopy, electron microscopy, chromatography or X-ray diffraction [7], where the information of biomolecules is extracted and stored in databases as protein data bank [8]. However, the employment of available data and CI techniques and models continues to be an area with many open questions, in terms of determining the best source of information, the election of using adequate features, or the application of best models [2, 9, 10].

In addition, machine learning (ML) models, which also can be considered as a subfield of artificial or computational intelligence, hold the capacity of learning from available data. In the particular case of proteins sequences and CI association, some researchers have tried to predict implications from mutations from amino acids sequences associated to an specific protein [11–13], a task that is not easily to determine. Furthermore, the use of ML models can provide information about the thermophilic properties [14], characterize and predict human aging [15], gene expression profile [17], and function prediction [16]. For the case of localization sites, there are different approaches that make use of information from subcellular analysis [17], subchloroplast [18], and sub-Golgi applications [19].

The proposed application was developed for a prokaryotic cell associated to *E. coli*, which is a Gram-negative bacterium, commonly present in the lower intestine of mammalians [20–22]. This bacterium is one of the most studied prokaryotic organism. *E. coli* proteome is obtained from 4288 protein-coding genes, where 38% have no attributed function. For this reason it is important to study the protein localization sites, which can contribute to function analysis.

In the present work, a comparison of the employment of two information sources from *E. coli* proteins was followed. The objective is to determine the subcellular localization sites of the macromolecule from features extracted of proteins for the two source cases: information from targeting signal and empirically sequence features [23]. Four different classes were established, according to the protein localizations sites: cytoplasm, periplasmic spaces, outer and inner membrane.

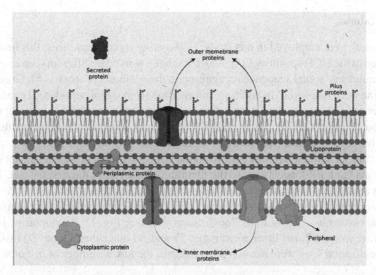

Fig. 1. Protein localization sites. Created with https://www.biorender.com/

2 Methodology

Figure 2 visualizes the block diagram for the employed methodology. Two sources of information were taken into account for determining the localization of proteins: targeting signal sequences and sequences solely. Then, three ML models were explored to find the best option for the location classification. Finally, an analysis of the features in terms of classification performance was implemented based on the best models. These aspects were studied for three possible scenarios: i) classification using the first one source, ii) classification employing the second information source, and iii) classification for the fusion of both sources.

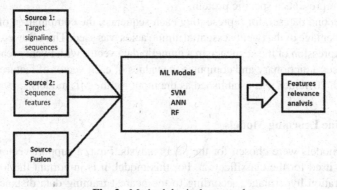

Fig. 2. Methodological proposal

2.1 Database

Two datasets were employed in this study: *i) Targeting signal sequences:* this first set is associated to the UCI repository [24], which has been worked in different scenarios [25]. There, features as signal sequence recognition in the synthesis process as McGoech and von Heijne were employed. In addition, information from signal peptidase II consensus, the presence of charge on the amino terminus (N-terminus), membrane spanning region prediction at using the computation of the number of transmembrane segments, and a score based on discriminant analysis of the amino acid content [25]. *ii) Sequence features:* the second dataset was built from the collection from protein sequences form protein data bank [26], according to the identifier employed in the first dataset. In this case, the amino acid sequences were used to obtain the features.

Finally, the after pre-processing stage, 291 out 335 data vectors were considered from four classes, according to the localization sites (see Fig. 1): *i)* cytoplasm: 142, *ii)* periplasmic space: 52, *iii)* inner membrane: 77 and *iv)* outer membrane: 20 instances. Other localization sites were not considered due to the lower number of instances.

2.2 Feature Extraction

First source of data, associated to targeting signal sequences, was not modified and it was not necessary a preprocessing step before the ML training. The included variables were: *mcg* - McGeoch's method for signal sequence recognition, *gvh* - von Heijne's method for signal sequence recognition, *lip* - von Heijne's Signal Peptidase II consensus sequence score, *chg* - presence of charge on N-terminus of predicted lipoproteins, *aac* - score of discriminant analysis of the amino acid content of outer membrane and periplasmic proteins, *alm1* - score of the ALOM membrane spanning region prediction program, and *alm2* - score of ALOM program after excluding putative cleavable signal regions from the sequence. The considered values were inside the [0–1] interval, with binary states for the signal peptidase II consensus and the presence of charge on N-terminus (*lip* and *chg*). In this case, a vector with seven values, according to the description in the previous subsection, represents a specific protein.

In the second dataset, for representing each sequence, the computation of presence through percentage of the twenty essential amino acids was used. This allowed to modify the string expression of the sequence in a numeric data vector with twenty features. The algorithm uses a dictionary and computes the values of each vector of features. Finally, a vector with 20 values was established as the input for the ML models.

2.3 Machine Learning Models

Three ML models were chosen for the data analysis. First, a support vector machine (SVM) was used for the classification. For this model, it is important the localization of the separation hyperplane, according to the nearest training data distance and the maximization of the margin of separation (C). For nonlinear separable classes, it is useful to apply a kernel to the training data [27]. The hyperparameters of the SVM were determined through an experimental mode. For this, the use of different options of kernel was explored, modifying possibilities as Gaussian, linear, radial basis function, polynomial and logistic. In addition, the C parameter was also tested with values between 0.1 and 1000. As the present application holds more than two classes, it was necessary to implement the strategy one versus all, where the SVM is used as an ensemble of classifiers.

The second model was based on an artificial neural network (ANN) founded on multilayer perceptron architecture. There, each input vector is projected on just one hidden layer [28]. The output layer was composed by four units, according to the protein localization sites. Different ANN architectures were studied, but modification of the number of units in the hidden layer, evaluating values in the interval five to thirty. Another explored hyperparameter was the option of using early stopping.

A third model known as random forest (RF) was evaluated. This forest is built through the exploration from different combinations of decision trees called estimators [29]. Processes of pruning and assignation of random sample from data are used in the training. The RF model obtains better performance in terms of generalization, due to a kind of compensation given by the trees proposals. The hyperparameters were obtained, employing an experimental method. The number of estimators was modified from ten to thirty and the criterion to measure the quantity of information was evaluated with Gini and Entropy indexes.

A cross-validation process with four folds was implemented, using the same sets for training and test for all models simultaneously. This number of folds was elected due to the dataset size, as the outer membrane class holds just twenty instances. According to a multiclass problem, the used metric was based on balance accuracy, where each class was weight, according to the number of instances, searching for a better performance for all classes. Fusion was carried out with a unique vector, joining information sources as input in the ML models.

The computational experimental setup was implemented using Python with libraries as Numpy, Pandas, and Scikit-learn. This last tool, with special capabilities for developing ML approaches, was used to determine the hyperparameters of the three models with the grid search strategy.

2.4 Features Relevance Analysis

In order to obtain information about the contribution of the features in each scenario, a relevance analysis was followed according to the classical approach provided in [30]. This process was implemented by replacing the original value of the input vector by a zeros vector.

For this, the best classifier from each scenario was taken into account. Then, all dataset was presented to the ML model with the modification in the input vectors and a performance was computed for each variable. Finally, the relevance of the feature was associated to low values of the classification performance. This means that when the variable was excluded in a zero vector manner, the classifier obtained worse results due to the relevance of this feature.

3 Results

Tables 1 and 2 visualize the results for both used sources in a cross-validation process. Table 3 exhibits the results when both sources were used in a simultaneous mode. At employing the four folds, it is possible to comprehend how the three scenarios obtained results with a balance accuracy upper than 89%, and how the use of first dataset produced the best results.

For the first case, where the input information is based on targeting signal sequences, the best model was the RF. This model was established by using between 20 and 30 estimators, the entropy criteria for computation of information in the internal nodes and a structure of five levels as maximum.

Table 1. Results for targeting signal sequences

Folds Number	ML Models		
	SVM	MLP	RF
1	0.95	0.92	0.95
2	0.91	0.91	0.95
3	0.83	0.85	0.86
4	0.87	0.87	0.90
Mean ± std	0.89 + 0.02	0.89 + 0.03	**0.92 + 0.04**

Table 2. Results for empirical sequences features

Folds Number	ML Models		
	SVM	MLP	RF
1	0.94	0.91	0.81
2	0.76	0.72	0.71
3	0.91	0.88	0.84
4	0.82	0.82	0.74
Mean ± std	**0.86 + 0.07**	0.83 + 0.07	0.77 + 0.05

Table 3. Results for fusion data

Folds Number	ML Models		
	SVM	MLP	RF
1	0.91	0.89	0.93
2	0.89	0.90	0.87
3	0.99	0.96	0.98
4	0.79	0.80	0.77
Mean ± std	**0.90 + 0.07**	0.89 + 0.06	0.89 + 0.08

In the second case, with information from amino acids sequence in the protein, the SVM was the model that exhibited the best performance. In most of fold, s the SVM employed a polynomial kernel and a C parameter of 1.

Finally, when the fusion of data information was used, the SVM was the best model with a 90% as balanced accuracy. For this model, each fold had a different C parameter, with values of 1, 10 and 100, but conserving the linear kernel. Figures 3, 4 and 5 show the confusion matrixes of the best results in each case.

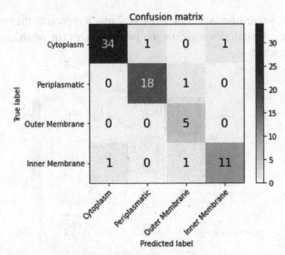

Fig. 3. Confusion matrix of the best result with RF model and features from targeting signal sequences.

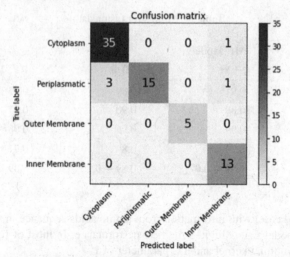

Fig. 4. Confusion matrix of the best result with SVM model and features from protein sequences.

Results for MLP models were obtained with between 15 and 20 units in the hidden layer. The fusion demanded a higher number of units in the hidden layer, requiring 30 neurons.

For the features relevance analysis, Figs. 6, 7 and 8 represent the results in each of the three scenarios. Visualizations show the performances for when each variable was replaced by a zeros vector.

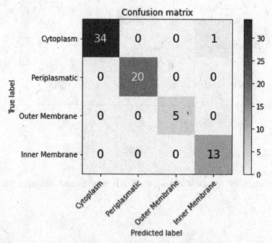

Fig. 5. Confusion matrix of the best result with SVM model and fusion of features from both datasets.

For the first source, it is seen how low performances were obtained when the unemployment of the *gvh* and *aac* variables was analyzed. These cases provided results with

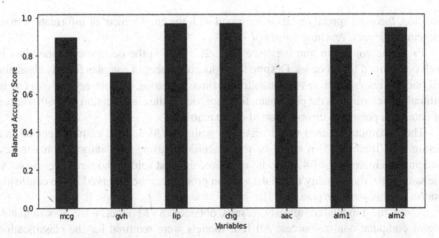

Fig. 6. Relevance analysis for the first source of information.

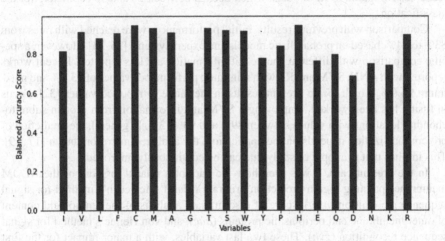

Fig. 7. Relevance analysis for the second source of information.

approximately 71% of accuracy for both variables and all dataset (see Fig. 6). The second scenario had impacted results when was affected the employment of information from G, Y, E, D, N and K amino acids with 75%, 77%, 76%, 68%, 74% y 74% respectively. Finally, the third scenario presented worse performances when *alm1*, *aac*, *gvh* and *mcg*, with accuracies of 45%, 65%, 52% and 68%, respectively.

4   Discussion

Through the reported results, it is possible to realize that there is not considerable difference when a fusion proposal was carried out. Results were diverse in the range from 77% of balanced accuracy to 92% in mean for the cross-validation with four folds. Best results were obtained with information from targeting signal sequences and the RF model. In addition, the most spread results were found with the employment of information from protein sequences, reaching values of 7%.

From the confusion matrixes (see Figs. 2, 3 and 4) the outer membrane class is well determined for all cases. Despite the reduced number of samples for this class, the ML models reached the classification for the three scenarios. As contrary, the class more difficult to determine is the periplasmatic space, due to the classification for both sources of data, as it presented errors reported in the matrixes.

The maximum reached result was 99% with a SVM and the employment of data fusion (see Table 3). This could be the potential of using both information sources, noticing the increment of 4% for the results of the best fold in the three scenarios. At the same time, the capacity of the data fusion process can be observed in the confusion matrix for this scenario (see Fig. 5).

About the models complexity, results obtained SVM models are less demanding of computational resources. All four models were required for the classification when the strategy one *versus* all was developed. This translates in less computational resources compared to RF models, which needed between 20 and 30 estimators for the classification.

Comparison with previous results, initial performances were reached with rates from 83% to 84%, based on probabilistic models and expert systems [25, 31]. However, a specific comparison with different sources of information are less reported. Recent works reported with ANN, SVM and RF with minimum performance values of 73.6% and maximum 96.6%, which locate present results in the range of reported values [32]. Focus in fusion has been worked, employing a SVM and information from protein submitochondria locations with values between 89% and 93% [33]. In general, the prediction of protein localization depends on the application, the analyzed micro-organism [17, 19]. This implies that is a topic of study that can be continuously improved.

In the present case, it was seen how the variables related to score of the ALOM membrane spanning region prediction program (*alm1*), McGeoch's method for signal sequence recognition (*mcg*), score of discriminant analysis of the amino acid content of outer membrane and periplasmic proteins (*aac*), and von Heijne's method for signal sequence recognition (*gvh*), These two last variables, with a major impact for the first scenario with just target signaling information, the performance achieved rates of 71% in accuracy. For the third scenario, the results for these variables did not have performances upper 70%, especially with values of 45% and 53% for the mentioned variables. The second scenario, when information from protein sequences was used, the most critical case was identified when the variable with data from aspartic acid was replaced by the zeros vector. In this case, the performance dropped to 68% in accuracy. Additionally, the glutamic acid had similar result for the third scenario, showing low performance in this E amino acid (see Fig. 8). Both amino acids with this effect in the accuracy result are known as the acidic ones, which can be related to the classification of protein localization in the present case.

Limitations of the present study can be related to the dataset size, where some classes as outer membrane held just 20 samples. This is a limitation in the training of ML models. However, in the present case, the classification for this localization was not affected for this aspect. For the information extracted from protein sequences, the size

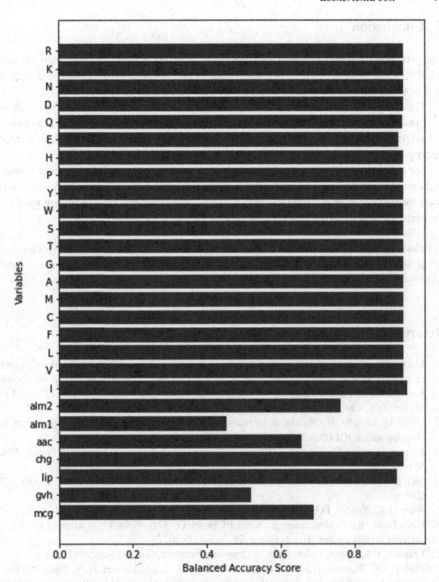

Fig. 8. Relevance analysis for the fusion of source of information.

of input vector with 20 features makes that ML models required more and dedicated attention in the training process. About this, the present work wanted to compared in a transparent mode both datasets, employing the same folds for training and test, models and its hyperparameters, for determining the contribution of using both and separate type of information.

5   Conclusion

The proposal of using a fusion of data from targeting signal sequence and protein sequences was analyzed. Three ML traditional models were trained for the classification of four localization sites for *Escherichia coli*.

These results contribute to the understanding on how information form amino acids in protein sequences provide small knowledge for the prediction of protein locatization sites in bacteria. In addition, the capacity of learning for the specific application demostrated better perfomance compared to the similar approaches in this field.

For the feature relevance analysis, the source with target signaling sequences information show a major importance for the classification. For fusion data sources, this result was more noteable, confirming the pertinence of incluiding this type of information for the classification treated in the present case.

Acknowledgment. Authors acknowledge the support of the Universidad del Rosario for funding this project. In addition, the contribution of research incubator teams *Semillero en Inteligencia Artificial en Salud: Semill-IAS* and *Semillero SyNERGIA*.

References

1. Martino, A., Giuliani, A., Rizzi, A.: Granular computing techniques for bioinformatics pattern recognition problems in non-metric spaces. In: Pedrycz, W., Chen, SM. (eds.) Computational Intelligence for Pattern Recognition. Studies in Computational Intelligence, vol. 777, pp. 53–81. Springer, Cham (2018). https://doi.org/10.1007/978-3-319-89629-8_3
2. Dash, S., Subudhi, B.: Handbook of research on computational intelligence applications in bioinformatics. IGI Global (2016)
3. Jumper, J., et al.: Highly accurate protein structure prediction with AlphaFold. Nature **596**, 583–589 (2021)
4. Wei, G.-W.: Protein structure prediction beyond AlphaFold. Nat. Mach. Intell. **1**, 336–337 (2019)
5. Auer, G.K., Weibel, D.B.: Bacterial cell mechanics. Biochemistry **56**, 3710–3724 (2017)
6. Nevo-Dinur, K., Govindarajan, S., Amster-Choder, O.: Subcellular localization of RNA and proteins in prokaryotes. Trends Genet. **28**, 314–322 (2012)
7. Branden, C.I., Tooze, J.: Introduction to protein structure. Garland Science (2012)
8. Burley, S.K., Berman, H.M., Kleywegt, G.J., Markley, J.L., Nakamura, H., Velankar, S.: Protein data bank (PDB): the single global macromolecular structure archive. Protein Crystallogr. 627–641 (2017)
9. Hassanien, A.E., Al-Shammari, E.T., Ghali, N.I.: Computational intelligence techniques in bioinformatics. Comput. Biol. Chem. **47**, 37–47 (2013)
10. Liu, M., Chen, X.: Computational intelligence and bioinformatics. Comput. Intell. **2**, 234 (2015)
11. Jamal, S., Khubaib, M., Gangwar, R., Grover, S., Grover, A., Hasnain, S.E.: Artificial Intelligence and Machine learning based prediction of resistant and susceptible mutations in Mycobacterium tuberculosis. Sci. Rep. **10**, 1–16 (2020)
12. Grønning, A.G.B., et al.: DeepCLIP: predicting the effect of mutations on protein–RNA binding with deep learning. Nucleic Acids Res. **48**, 7099–7118 (2020)

13. Orjuela-Cañón, A.D., Figueroa-García, J.C., Neruda, R.: Automated machine learning strategies to damage identification of neurofibromatosis mutations. In: 2021 20th IEEE International Conference on Machine Learning and Applications (ICMLA), pp. 1341–1344 (2021)
14. Wang, X.-F., Gao, P., Liu, Y.-F., Li, H.-F., Lu, F.: Predicting thermophilic proteins by machine learning. Curr. Bioinform. **15**, 493–502 (2020)
15. Kerepesi, C., Daróczy, B., Sturm, Á., Vellai, T., Benczúr, A.: Prediction and characterization of human ageing-related proteins by using machine learning. Sci. Rep. **8**, 1–13 (2018)
16. Bonetta, R., Valentino, G.: Machine learning techniques for protein function prediction. Proteins Struct. Funct. Bioinform. **88**, 397–413 (2020)
17. Wan, S., Mak, M.-W.: Machine learning for protein subcellular localization prediction. In: Machine Learning for Protein Subcellular Localization Prediction. De Gruyter (2015)
18. Liu, M.-L., et al.: An overview on predicting protein subchloroplast localization by using machine learning methods. Curr. Protein Pept. Sci. **21**, 1229–1241 (2020)
19. Yang, W., Zhu, X.-J., Huang, J., Ding, H., Lin, H.: A brief survey of machine learning methods in protein sub-Golgi localization. Curr. Bioinform. **14**, 234–240 (2019)
20. Vila, J., et al.: Escherichia coli: an old friend with new tidings. FEMS Microbiol. Rev. **40**, 437–463 (2016)
21. Keseler, I.M., et al.: EcoCyc: a comprehensive database of Escherichia coli biology. Nucleic Acids Res. **39**, D583–D590 (2010)
22. Allocati, N., Masulli, M., Alexeyev, M.F., Di Ilio, C.: Escherichia coli in Europe: an overview. Int. J. Environ. Res. Public Health. **10**, 6235–6254 (2013)
23. Yu, C.-S., Chen, Y.-C., Lu, C.-H., Hwang, J.-K.: Prediction of protein subcellular localization. Proteins Struct. Funct. Bioinform. **64**, 643–651 (2006)
24. Dua, D., Graff, C.: {UCI} Machine Learning Repository (2017). http://archive.ics.uci.edu/ml
25. Nakai, K., Kanehisa, M.: Expert system for predicting protein localization sites in gram-negative bacteria. Proteins Struct. Funct. Bioinform. **11**, 95–110 (1991)
26. Abola, E.E., Bernstein, F.C., Koetzle, T.F.: The protein data bank. In: Schoenborn, B.P. (eds.) Neutrons in Biology. Basic Life Sciences, vol. 27, p. 441. Springer, Boston, MA (1984). https://doi.org/10.1007/978-1-4899-0375-4_26
27. Anam, K., Al-Jumaily, A.: Evaluation of extreme learning machine for classification of individual and combined finger movements using electromyography on amputees and non-amputees. Neural Netw. **85**, 51–68 (2017)
28. Haykin, S.: Neural Networks and Learning Machines. Pearson, London (2009)
29. Bergstra, J., Bengio, Y.: Random search for hyper-parameter optimization. J. Mach. Learn. Res. **13** (2012)
30. Seixas, J.M., Calôba, L.P., Delpino, I.: Relevance criteria for variable selection in classifier designs. In: International Conference on Engineering Applications of Neural Networks, pp. 451–454 (1996)
31. Horton, P., Nakai, K.: A probabilistic classification system for predicting the cellular localization sites of proteins. In: Ismb, pp. 109–115 (1996)
32. Tiwari, A.K., Srivastava, R.: A survey of computational intelligence techniques in protein function prediction. Int. J. Proteomics 2014 (2014)
33. Zakeri, P., Moshiri, B., Sadeghi, M.: Prediction of protein submitochondria locations based on data fusion of various features of sequences. J. Theor. Biol. **269**, 208–216 (2011)

Artificial Bee Colony-Based Dynamic Sliding Mode Controller for Integrating Processes with Inverse Response and Deadtime

Jorge Espin[1,2] , Sebastian Estrada[1,2] , Diego S. Benítez[2(✉)] ,
and Oscar Camacho[2]

[1] Department of Electronics Engineering, Universidad Técnica Federico Santa María,
Valparaíso, Chile
{jorge.espin,juan.estradac}@usm.cl
[2] Colegio de Ciencias e Ingenierías "El Politécnico", Universidad San Francisco de
Quito USFQ, Quito 170157, Ecuador
{dbenitez,ocamacho}@usfq.edu.ec

Abstract. A strategy for optimizing the settings of a dynamical sliding mode controller using an artificial bee colony optimization algorithm is proposed in this paper. The performance of the obtained controller is then evaluated and compared to that of a conventional PID and a dynamical sliding mode controller that has been optimized through a heuristics-based strategy by simulating two integrating linear systems with dead time and inverse response. By utilizing the suggested strategy, it was possible to increase both performance indices and transient characteristics.

Keywords: Dynamic sliding mode control · Artificial bee colony · Integrating systems · Dead time · Inverse response

1 Introduction

Dynamic Sliding Mode Controller (DSMC) is a control strategy that adds additional dynamics that act as compensators to achieve and improve system stability, obtaining desired system behavior and performance. Therefore, an option to reduce the chattering effect is DSMCs, which have the advantage of not decreasing performance on the controller because no smoothing functions are used. This feature is appropriate for chemical processes control where the discontinuities cannot be allowed on the final element of control and fast vibrations are not tolerated due to the adverse effects on the quality of the product [32].

In the literature, several works are associated with the application of DSMC to different kinds of processes. For example, in [33] this controller was effectively used in chemical reactors, and in [8] to wind-driven induction generators. A DSMC approach from a First-Order plus Deadtime (FOPDT) model was developed in [29] to obtain a simple controller with a fixed structure and tested in a high delay system. While [5] and [3] proposed a dynamical control scheme combining concepts of Iinoya's compensator and sliding mode control for inverse response process, and [13] introduced a DSMC founded on the Iinoya and Alpeter proposal and the Sliding Mode Control (SMC) design procedure. This approach was successfully applied to high-order chemical processes with an inverse condition and high delay. The controller performance was tested by simulation of a nonlinear mixing tank with variable dead time and its implementation on an Arduino Temperature Control Lab.

However, most of the work associated with tuning parameters is for SMCs. For example, Rojas et al. [31] implemented it for open-loop unstable systems. While [25] presented a chattering-free SMC for a robot manipulator, which includes a proportional-integral-derivative (PID) part with a fuzzy tunable gain. Piltan et al. [28] used a switching function in the presence of a mathematical error-based method instead of a switching function method in a pure SMC to reduce the chattering. They applied a mathematical model-free tuning method to the SMC for adjusting the sliding surface gain (λ). [26] used Genetic Algorithms (GA) to study the problem of designing a stable sliding mode that yields robust performance in variable structure control systems, showing that for various instances, the GA is viable and has great potential in the design of SMC systems. Mehta and Rojas [24] used the cuckoo optimization algorithm [30] to provide a tuning method for a Smith Predictor based on sliding mode for unstable systems. Their approach reduces the Integral Squared Timer Error (ISTE) and controller signal variability.

In such a sense, [4] suggested an optimization procedure to obtain the tuning parameters. The optimization criterion considers the Integral Square Error (ISE), the settling time, and the overshoot as restrictions. The results are then applied to the process with high delay. The simultaneous tuning of numerous parameters can be considered a real challenge if not have enough background about the process, which leads to taking this issue as a continuous optimization problem [17,38]. An Alternative for finding the values of the parameters can be to use optimization algorithms based on swarm intelligence. Most of these well-established optimization techniques study and model the behavior of interacting populations agents or swarms and their ability to organize themselves [17], such as an ant colony [10], a flock of birds, a school of fish, or an immune system [9]. There are other optimization alternatives widely used in the literature, such as GA [14] and Particle Swarm Optimization (PSO) [19]. Although the optimization methods mentioned above have yielded promising results for tuning problems, in this paper, we explore using a bee swarm-based optimization algorithm called Artificial Bee Colony (ABC) [1,2,16,17] for this problem. We choose ABC optimization based on the comparisons made in [17], where the results obtained from ABC outperform the other optimization algorithms.

Furthermore, integrating processes have no equilibrium state, such as gas pressure and liquid level systems. A classic example of a nonself-regulating process with an inverse response is the level control of a boiler steam drum [20]. Closed-loop control of integrating processes is an exciting and challenging subject in control engineering. Unfortunately, adding right-hand zeros to the transfer function of the process causes an inverse response and makes the control more difficult. Previously, in [11], we designed a DSMC for integrating systems with an inverse response and dead time based on a reduced process model, where the tuning of the controller was done in a heuristic way, obtaining an adequate performance. However, to improve the performance of such work, this paper proposes using ABC as an alternative to enhance and get better tuning of the parameters of the controller. The proposal is then compared against the original approach by simulation in two integrating linear systems with dead time and inverse response.

This article is summarized as follows: Sect. 2 explains the fundamental concepts. Section 3 describes the tuning of the DSMC using the ABC algorithm. Section 4 provides the experimentation and simulation results comparing multiple controllers. Finally, Sect. 5 outlines the conclusions drawn from this paper.

2   Fundamentals

2.1   Sliding Mode Control Foundations

SMC theory is derived from Variable Structure Control (VSC) studied by Utkin [34]. SMC is considered as a nonlinear feedback control strategy with a high-frequency switching control action [21,34,36]. In addition, SMC exhibits remarkable features for controlling highly nonlinear systems, rejecting external disturbances and being insensitivity to parameter variations [21,36,37].

SMC aims to drive the initial condition onto a predefined sliding surface. Once the surface is attainable, the state will slide over until the desired state is reached. Hence, SMC is a two-part controller, the first one is related to sliding mode, and the other one is concerned with the sliding surface reachability [7,13, 36]. The SMC law is described as

$$U(t) = U_C(t) + U_D(t) \qquad (1)$$

where,

- $U(t)$: SMC law
- $U_C(t)$: Continuous function
- $U_D(t)$: Discontinuous function

The main undesirable phenomenon of SMC theory is known as "chattering", which has oscillations with finite frequency and amplitude. Chattering is a highly damaging phenomenon since it causes degradation in the control system accuracy. Furthermore, due to high-frequency oscillations in the control action, it is very likely to wear mechanical parts in actuators and result in heat losses in power circuits [13,35,36]. This negative effect is produced by the SMC discontinuous function details in (2).

$$U_D(t) = K_d sign(S(t)) \qquad (2)$$

To minimize and to smooth the chattering effect, the expression $sign(S(t))$ is approximated to a sigmoid function as recommended in [7,13]; yielding (3).

$$U_D(t) = K_d \frac{S(t)}{|S(t)| + \delta} \tag{3}$$

K_d and δ are tuning parameters.

2.2   Dynamic Sliding Mode Control

DSMC is a control technique with characteristics of fast response and stabilization, robustness, and low sensitivity to modeling errors, becoming a subject of study in recent years [3,6,11,13,22]. In addition, its main objective is to eliminate the chattering effect caused by the SMC discontinuous function [13]. Hence, DSMC incorporates additional dynamics to transform the discontinuous function and avoid the use of smoothing functions without decreasing the controller performance [6,32].

Likewise, DSMC has two main components, the continuous function, which is closely related to the sliding mode, and the discontinuous function concerning the surface reachability, as shown in (4).

$$\dot{U}(t) = \dot{U}_C(t) + \dot{U}_D(t) \tag{4}$$

where,

- $\dot{U}(t)$: DSMC Law
- $\dot{U}_C(t)$: Continuous function
- $\dot{U}_D(t)$: Discontinuous function.

2.3   Artificial Bee Colony Optimization Algorithm

The ABC algorithm was proposed by Karaboga [16] and is regarded as a swarm intelligence iterative method that inspects the foraging behavior of honey bee swarms and how they share information in the hive [17]. It is commonly used to solve numerical, combinatorial, and multidimensional optimization problems [2]. The ABC algorithm is inspired by the foraging behavior of honey bee swarm. This biological animal society comprises three types of honeybees: employed, onlooker, and scout bees. Each has the task of finding the best food source location with a sufficient nectar amount [17]. The flowchart of Fig. 1 shows the different algorithm phases according to the types of honeybees. Once the food source is located in ABC, its quality is evaluated. Therefore, it could be assumed that the food source location represents the final best solution, and the nectar amount corresponds to the quality of the solution [12].

The honeybees foraging behavior determines a foraging space. The initialization stage selects random food source positions by the employed bees. When these positions are located, the employed bees move to them and obtain information about location and nectar amount. With these defined values, they return

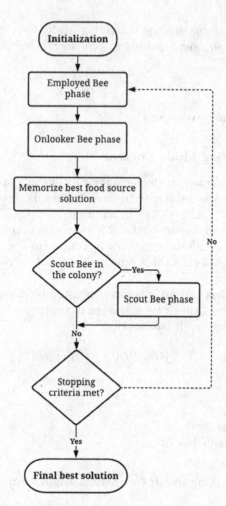

Fig. 1. ABC algorithm general flowchart (adapted from [12]).

to the hive and share this information with the onlookers' bees, which define the food source quality. Therefore, the quantity of employed and onlooker bees must equal the food sources. On the other hand, the scout bees must discover new locations when the food sources are exhausted [2,12,16].

3 Dynamical Sliding Mode Control Tuning Using ABC

The Dynamic Sliding Mode Controller for Inverse Response Integrating Systems (DSMC-IRIS) with dead time developed by Espin and Camacho in [11], is a novel control alternative for complex dynamic systems due to its robustness and low susceptibility to modeling mismatches.

The controller design uses a PID sliding surface as described in (5).

$$S(t) = \lambda_1 e(t) + \lambda_0 \int_0^t e(t)\, dt + \frac{de(t)}{dt} \tag{5}$$

The reduced-order inverse-response integrating system model with dead time is detailed in (6).

$$G_p(s) = K\frac{-\eta s + 1}{s(\tau s + 1)}e^{-t_0 s} \tag{6}$$

The model-based DSMC-IRIS [11] is synthesized and then, the full controller expression is obtained in (7).

$$\dot{M}(t) = \frac{1}{K(\lambda - \eta)}[(1 - \lambda_1)\dot{Y}^*(t) + \tau\lambda_0 e(t) - KM(t)]...$$

$$... + K_d sign(S(t)) \tag{7}$$

The hybrid control scheme shown in Fig. 2 has a classical PD controller in the inner loop, defined by:

$$PD = K_o(T_d s + 1) \tag{8}$$

$G_m(s)$ corresponds to the Smith predictor fast model,

$$G_m(s) = K\frac{-\eta s + 1}{s(\tau s + 1)} \tag{9}$$

and $G_i(s)$ is related to the Iinoya and Alpeter compensator [15], responsible for mitigating the inverse condition,

$$G_i(s) = K\frac{\lambda s}{s(\tau s + 1)} \tag{10}$$

Previous work presents a trial-and-error tuning of the following parameters λ_1, λ_0, K_d, K_o, T_d, and λ.

3.1 Lyapunov Stability Condition

The following inequality is considered for the Lyapunov stability condition:

$$\dot{V} = S\dot{S} < 0 \tag{11}$$

$\dot{S}(t)$ is described by:

$$\frac{dS(t)}{dt} = -\frac{d^2Y^*(t)}{dt^2} - \lambda_1\frac{dY^*(t)}{dt} + \lambda_0 e(t) \tag{12}$$

DSMC-IRIS bases its design on the SP fast model using the Iinoya and Alpeter approach and the difference between process response and SP output. Yielding the following:

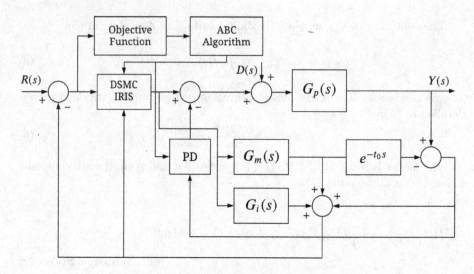

Fig. 2. DSMC-IRIS-ABC control scheme proposal

$$\frac{Y^*(s)}{M(s)} = K\frac{[(\lambda - \eta)s + 1]}{s(\tau s + 1)} \tag{13}$$

(13) can be expressed in differential equation form as shown in (14)

$$\frac{\tau d^2 Y^*(t)}{dt^2} + \frac{dY^*(t)}{dt} = K(\lambda - \eta)\frac{dM(t)}{dt} + KM(t) \tag{14}$$

Substituting DSMC-IRIS's control law of (7) in (14) and solving its highest derivative yields:

$$\frac{d^2 Y^*(t)}{dt} = -\lambda_1 \frac{dY^*(t)}{dt} + \lambda_0 e(t) + \frac{K(\lambda - \eta)K_d}{\tau} sign(S(t)) \tag{15}$$

Replacing (15) in (12), the following is obtained:

$$\frac{dS(t)}{dt} = -\frac{K(\lambda - \eta)K_d}{\tau} sign(S(t)) \tag{16}$$

Therefore, the reaching condition is defined by:

$$S\dot{S} = -\frac{K(\lambda - \eta)K_d}{\tau} S(t)sign(S(t)) < 0 \tag{17}$$

$$-\frac{K(\lambda - \eta)K_d}{\tau}|S(t)| < 0$$

where,

$$\mathbb{K} = \frac{K(\lambda - \eta)K_d}{\tau} \tag{18}$$

So,

$$- \mathbb{K}|S(t)| < 0 \tag{19}$$

Finally, to make sure the stability of the controller, it is necessary to guarantee that $\mathbb{K} > 0$.

3.2 Objective Function

The parameter solutions were found using the ABC optimization algorithm, where the objective function J to be minimized is given by the Integral of Squared Error (ISE), as shown in (20).

$$J = ISE = \int_0^t e^2(t)dt \tag{20}$$

4 Experimentation and Simulation Results

This section analyzes and compares previous approaches concerning the ABC tuning optimization proposal to control two linear integrating systems with an inverse response and dead time. The quantitative analysis is based on performance indexes (ISE and Total Variation of control effort (TVu)) and transient characteristics (Settling time (Ts) and Overshoot (OS %)). The ABC algorithm source code for Matlab was obtained from [1]; for the simulation and the ABC source code execution, a 3.1 GHz Dual-Core Intel Core i7 PC with 16 GB of RAM and a macOS operating system was used.

4.1 High-Order Linear Integrating System with Inverse Response and Dead Time

Example 1: The high-order linear integrating system with an inverse response and dead time shown in (21) and studied by Pai et al. in [27].

$$G_p(s) = \frac{0.5(-0.5s + 1)}{s(0.5s + 1)(0.4s + 1)(0.1s + 1)}e^{-0.7s} \tag{21}$$

The synthesis and tuning of the controller require an integrating reduced-order model with an inverse response and dead time, so we used a computational characterization method proposed by Luyben in [23]. Obtaining,

$$G_{pm}(s) = \frac{0.583(-0.4699s + 1)}{s(1.1609s + 1)}e^{-0.81s} \tag{22}$$

To control the high-order linear integrating system with inverse response and dead time, Kaya declares the following tuning parameters [18]: $K_c = 0.87$, $T_i = 24$, $K_f = 0.0777$, $T_f = 0.0194$, $K_d = 0.74$, and $T_d = \tau = 1.1609$.

DSMC-IRIS [11] and DSMC-IRIS-ABC have the next tuning values, as shown in Table 1.

Table 1. Tuning parameters for Example 1

Controller	λ_1	λ_0	K_d	K_o	T_d	λ
DSMC-IRIS	2.096	1.098	1.294	0.709	0.788	3.000
DSMC-IRIS-ABC	1.308	1.516	1.929	0.800	0.583	3.112

To obtain DSMC-IRIS-ABC values, we used a colony size of $N_p = 20$, the number of the onlooker and employed bees corresponded to the half of colony size, respectively, and the number of cycles was five trials in the ABC algorithm. Its execution time was 968.80 s.

The responses of the system to the tracking test and load disturbance rejection of -0.5 amplitude at t $= 30$ s are found in Fig. 3. These results indicate that the proposed hybrid control topology enhances the transient characteristics of the high-order linear integrating system with an inverse response and dead time; it also provides a faster response to load disturbances than the other controllers. DSMC-IRIS-ABC did not initially exhibit an oscillatory response when we introduced a step change. Moreover, it quickly adjusts to the desired state after rejecting the load disturbance at 30 s. Example 1 requires proven robust controllers due to their dynamics, so the controller based on the dynamic sliding mode theory is a promising alternative compared to the traditional ones. The ABC algorithm made it possible to quickly find the previous DSMC-IRIS tuning parameters with better performance results.

The control actions are shown in Fig. 4 and exhibit similar behavior. However, DSMC-IRIS-ABC is saturated in the range of [0–1], while the other controllers exceed these values, causing more significant stress on the final control element. Although the proposed controller corrects the load disturbance quickly, it also verifies a more significant effort in the control action. A comparative quantitative analysis is detailed in the bar chart of Fig. 5. The DSMC-IRIS-ABC controller has the best transient characteristics and performance indexes since it has the lowest magnitude values, generating better advantages for systems with this dynamics type. On the other hand, the PID array proposed by Kaya is the control scheme with the worst performance, which indicates that traditional controllers are not a good option for systems with complex dynamics. Furthermore, it is found that tuning using the ABC algorithm significantly improves the DSMC-IRIS performance. An improvement of 16.49% in ISE, 2.8% in TVu, 1.141 s in Ts, and 2% in OS % is determined.

4.2 Linear Integrating Inverse Response System with Long Dead Time and Time Constant

Example 2: The linear integrating inverse response system with long-dead time and time constant shown in (23) and considered by Kaya in [18].

$$G_p(s) = \frac{(-s+1)}{s(10s+1)}e^{-7s} \tag{23}$$

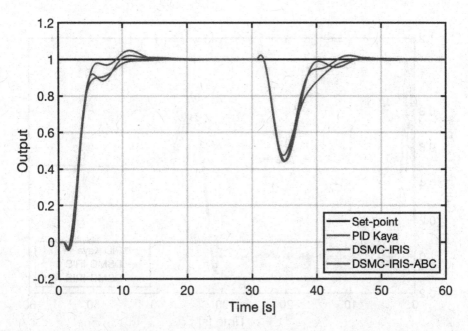

Fig. 3. Comparison between controllers responses with a load disturbance at t = 30 s for Example 1.

Kaya states the following tuning parameters: $K_c = 0.25$, $T_i = 2$, $T_f = 2.6734$, $K_f = 0.8863$, $T_d = T = 7.00$ and $K_d = 0.0624$.

DSMC-IRIS [11] and DSMC-IRIS-ABC have the next tuning values, as shown in Table 2.

Table 2. Tuning parameters for Example 2

Controller	λ_1	λ_0	K_d	K_o	T_d	λ
DSMC-IRIS	0.243	0.015	5.608	0.080	7	15
DSMC-IRIS-ABC	0.436	0.066	7.454	0.079	6.568	8.95

To obtain DSMC-IRIS-ABC values, we used a colony size of $N_p = 15$, the number of the onlooker and employed bees corresponded to the half of colony size, respectively, and the number of cycles was 20 trials in the ABC algorithm. Its execution time was 3820.14 s.

Figure 6 shows the responses of the system to tracking test and load disturbance rejection of −0.1 at t = 250 s.

Comparing DSMC-IRIS-ABC with the controller previously designed in [11] and tuned by trial and error, there is an improvement in the settling time. This change is produced by an increase in the K_d constant of 1.846 units plus

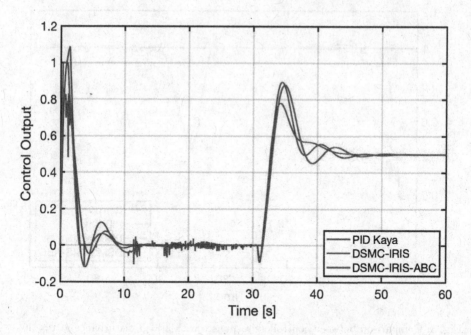

Fig. 4. Comparison of the output of the three controllers of Example 1.

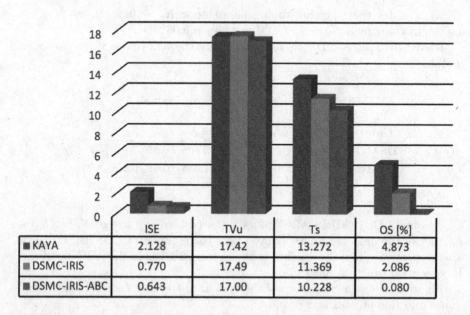

	ISE	TVu	Ts	OS [%]
■ KAYA	2.128	17.42	13.272	4.873
■ DSMC-IRIS	0.770	17.49	11.369	2.086
■ DSMC-IRIS-ABC	0.643	17.00	10.228	0.080

Fig. 5. Comparison of the performance indexes and transient characteristics of the three controllers of Example 1.

the lambda constant reduces its magnitude by 6.05 units. These two significant changes cause a faster response of the proposed controller. Similarly, one of the features that DSMC-IRIS-ABC stands out concerning the PID array proposed by Kaya [18] is the absence of overshoot. The proposal maintains a fast response and easily adjusts to the desired state.

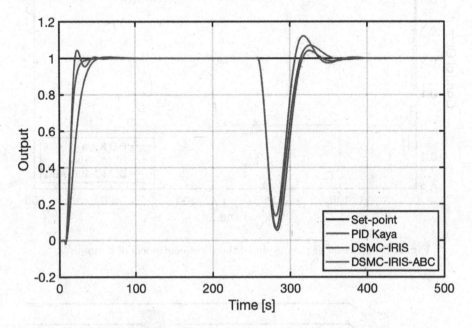

Fig. 6. Comparison of the responses of the three controllers of Example 2 for a load disturbance at t = 250 s.

Figure 7 shows the controllers outputs. Compared to the previous work, the proposed DSMC-IRIS-ABC has a high effort when detecting the step change. Faster response and no overshoot compensate for this drawback. Concerning the PID array, the proposal has a smoother control action causing less stress in the final control element than a traditional PID. While analyzing the load disturbance rejection at 250 s, there is no difference between the involved controllers.

A comparative quantitative analysis is detailed in the bar chart of Fig. 8. Where a better performance of DSMC-IRIS-ABC compared to the other controllers is observed, the proposal of this work has the best performance indexes and has better transient characteristics for the long-dead time and time constant system. An improvement of 45.91% in ISE, 3.62% in TVu, 14.26 s in Ts, and 0.68% in OS % is perceived compared to DSMC-IRIS.

4.3 Parameters Uncertainty Test

In this regard, a high-order linear integrating system with an inverse response and dead time is subjected to the variation of a parameter to generate a mismatch

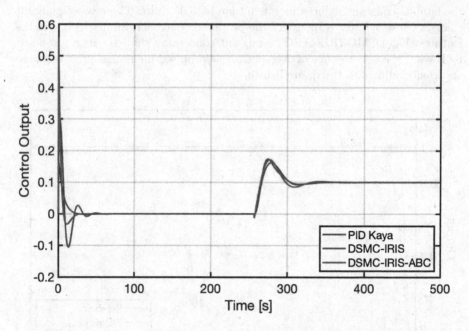

Fig. 7. Comparison of the output of the three controllers of Example 2.

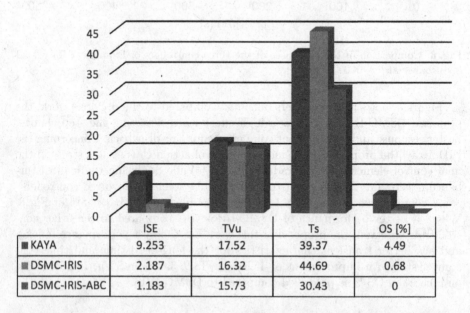

	ISE	TVu	Ts	OS [%]
■ KAYA	9.253	17.52	39.37	4.49
■ DSMC-IRIS	2.187	16.32	44.69	0.68
■ DSMC-IRIS-ABC	1.183	15.73	30.43	0

Fig. 8. Comparison of the performance indexes and transient characteristics of the three controllers of Example 2.

in the process reduced-order model expressed in (22). This alteration can be expressed with the increase or decrease of the actual value of time constant and dead time (τ, t_0). The above control schemes are based on the internal model, so this uncertainty is extremely demanding for controllers with this feature. The results are described below.

Example 1: Figures 9 and 10 depict the effect caused by the variation of a parameter in time constant and dead time of the high-order linear integrating system with an inverse response and dead time presented by Pai et al. in [27]. The results consistently describe an internal alteration in controller performance. Kaya [18] and DSMC-IRIS [11] suffer to a great extent the consequences from this uncertainty. Nevertheless, ABC-based DSMC-IRIS maintains the desired condition and attains the reference with a lower steady-state error. The proposal remains within an acceptable error range in case of dealing with similar variability. Likewise, if we analyze the behavior after introducing the load disturbance at 30 s, DSMC-IRIS and Kaya show an oscillatory behavior until they reach the final desired value. Conversely, this approach suggests a normal behavior and does not exhibit any oscillatory condition. This feature is emphasized in Fig. 10, in which both controllers strongly exhibit these oscillations caused by modeling mismatches.

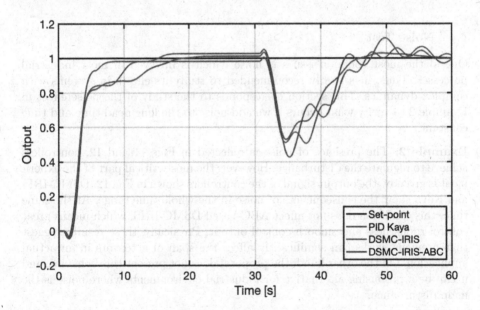

Fig. 9. Comparison of the responses of the three controllers of Example 1 for parameters uncertainty test and load disturbance at 30 s.

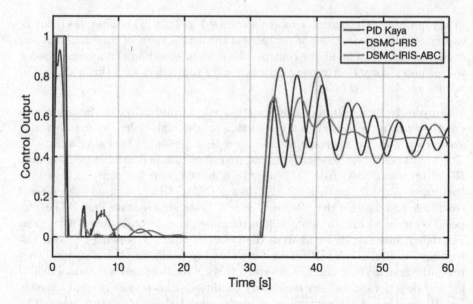

Fig. 10. Comparison of the output of the three controllers of Example 1 for parameters uncertainty test and load disturbance at 30 s.

4.4 Noise Test

One of the most common issues is noise which is involved in most industrial processes. Thus, it is highly recommended to study its effects in systems with complex dynamics. This section corresponds to the study of the noise effect in Example 2 to verify what occurs if we add noise to the long dead time and time constant.

Example 2: The presence of noise is reflected in Figs. 11 and 12, controllers achieve to regulate the disturbance. However, the most critical part of the experiment is given by the output signal of the controllers shown in Fig. 12. DSMC-IRIS and Kaya show the consequences of noise in the whole time range. At the same time, this condition does not affect ABC-based DSMC-IRIS, which means great control robustness and smooth control output. Physically, these results suggest that previous works can significantly affect the span of actuators in an actual application. On the other hand, the proposal does not present this behavior, and it can be a promising alternative for industrial environments where noise is the main disturbance.

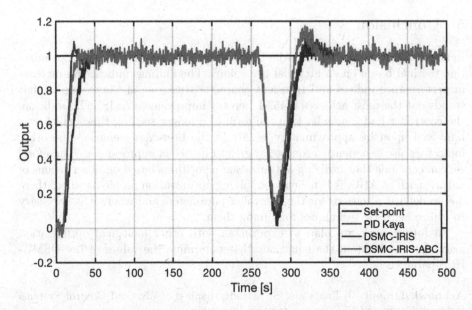

Fig. 11. Comparison of the responses of the three controllers of Example 2 for noise test and load disturbance at 250 s.

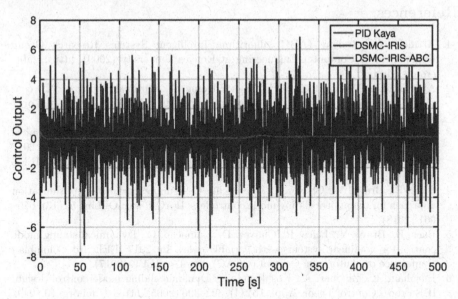

Fig. 12. Comparison of the output of the three controllers of Example 2 for noise test and load disturbance at 250 s.

5   Conclusion

In this study, the parameters of the DSMC-IRIS are tuned by using an optimization method based on an artificial bee colony. The findings indicate an increase in performance indices and transient characteristics; one of the examples that stands out the most achieves a 45.91 percent improvement in ISE. In addition, the controller had a drawback in the form of a longer settling time, which was improved upon by approximately 14.26 s in the best-case scenario. The other indicators also experience a somewhat greater degree of improvement. Therefore, we can conclude that utilizing optimization algorithms based on observations of nature, such as ABC, is beneficial for solving optimization problems where there are no defined equations for the controller parameters and where it is necessary to utilize heuristic techniques for tuning them.

In future work, we plan to experiment with other biological optimization algorithms and obtain the equations that determine the values of the DSMC-IRIS parameters.

Acknowledgment. J. Espin and S. Estrada thank the Advanced Control Systems Research Group at USFQ for the research internship.

References

1. Artificial Bee Colony (ABC) algorithm. Intelligent Systems Research Group Department of Computer Engineering at Erciyes University (2009). https://abc.erciyes.edu.tr/
2. Abachizadeh, M., Yazdi, M.R.H., Yousefi-Koma, A.: Optimal tuning of PID controllers using artificial bee colony algorithm, pp. 379–384 (2010). https://doi.org/10.1109/AIM.2010.5695861
3. Asimbaya, E., Cabrera, H., Camacho, O., Chávez, D., Leica, P.: A dynamical discontinuous control approach for inverse response chemical processes. In: 2017 IEEE 3rd Colombian Conference on Automatic Control (CCAC), pp. 1–6. IEEE (2017)
4. Baez, E., Bravo, Y., Chavez, D., Camacho, O.: Tuning parameters optimization approach for dynamical sliding mode controllers. IFAC-PapersOnLine 51(13), 656–661 (2018)
5. Báez, E., Bravo, Y., Leica, P., Chávez, D., Camacho, O.: Dynamical sliding mode control for nonlinear systems with variable delay. In: 2017 IEEE 3rd Colombian Conference on Automatic Control (CCAC), pp. 1–6. IEEE (2017)
6. Burnham, K., Zinober, A., Koshkoùei, A.: Dynamic sliding mode control design. IEE Proc. - Control Theor. Appl. 152(4), 392–396 (2005). https://doi.org/10.1049/ip-cta:20055133
7. Camacho, O., Smith, C.A.: Sliding mode control: an approach to regulate nonlinear chemical processes. ISA Trans. 39(2), 205–218 (2000). https://doi.org/10.1016/S0019-0578(99)00043-9
8. De Battista, H., Mantz, R.J., Christiansen, C.F.: Dynamical sliding mode power control of wind driven induction generators. IEEE Trans. Energy Convers. 15(4), 451–457 (2000)

9. De Castro, L., José, F., von Zuben, A.A.: Artificial immune systems: Part I-basic theory and applications (2000)

10. Dorigo, M., Birattari, M., Stutzle, T.: Ant colony optimization. IEEE Comput. Intell. Mag. **1**(4), 28–39 (2006). https://doi.org/10.1109/MCI.2006.329691

11. Espín, J., Camacho, O.: A proposal of dynamic sliding mode controller for integrating processes with inverse response and deadtime. In: 2021 IEEE Fifth Ecuador Technical Chapters Meeting (ETCM), pp. 1–6. IEEE (2021)

12. Ghanem, W., Jantan, A.: Using hybrid artificial bee colony algorithm and particle swarm optimization for training feed-forward neural networks. J. Theor. Appl. Inf. Technol. **67**, 664–674 (2014)

13. Herrera, M., Camacho, O., Leiva, H., Smith, C.: An approach of dynamic sliding mode control for chemical processes. J. Process Control **85**, 112–120 (2020)

14. Holland, J.H.: Adaptation in Natural and Artificial Systems. University of Michigan Press, Ann Arbor (1975). Second edition, 1992

15. Iinoya, K., Altpeter, R.J.: Inverse response in process control. Ind. Eng. Chem. **54**(7), 39–43 (1962)

16. Karaboga, D.: An idea based on honey bee swarm for numerical optimization. Technical report - tr06. Erciyes University (2005)

17. Karaboga, D., Basturk, B.: A powerful and efficient algorithm for numerical function optimization: artificial bee colony (ABC) algorithm. J. Glob. Optim. **39**, 459–471 (2007). https://doi.org/10.1007/s10898-007-9149-x

18. Kaya, Ibrahim: Controller design for integrating processes with inverse response and dead time based on standard forms. Electr. Eng. **100**(3), 2011–2022 (2018). https://doi.org/10.1007/s00202-018-0679-7

19. Kennedy, J., Eberhart, R.: Particle swarm optimization. In: Proceedings of ICNN'95 - International Conference on Neural Networks, vol. 4, pp. 1942–1948 (1995). https://doi.org/10.1109/ICNN.1995.488968

20. Korupu, V.L., Muthukumarasamy, M.: A comparative study of various smith predictor configurations for industrial delay processes. Chem. Prod. Process Model. **17**(6), 701–732 (2021)

21. Kunusch, C., Puleston, P., Mayosky, M.: Fundamentals of sliding-mode control design. In: Kunusch, C., Puleston, P., Mayosky, M. (eds.) Sliding-Mode Control of PEM Fuel Cells. Advances in Industrial Control, pp. 35–71. Springer, London (2012). https://doi.org/10.1007/978-1-4471-2431-3_3

22. Lin, F.J., Chen, S.Y., Shyu, K.K.: Robust dynamic sliding-mode control using adaptive RENN for magnetic levitation system. IEEE Trans. Neural Netw. **20**(6), 938–951 (2009). https://doi.org/10.1109/TNN.2009.2014228

23. Luyben, W.L.: Identification and tuning of integrating processes with deadtime and inverse response. Ind. Eng. Chem. Res. **42**(13), 3030–3035 (2003)

24. Mehta, U., Rojas, R.: Smith predictor based sliding mode control for a class of unstable processes. Trans. Inst. Meas. Control. **39**(5), 706–714 (2017)

25. Mohammad, A., Ehsan, S.S.: Sliding mode PID-controller design for robot manipulators by using fuzzy tuning approach. In: 2008 27th Chinese Control Conference, pp. 170–174. IEEE (2008)

26. Moin, N., Zinober, A., Harley, P.: Sliding mode control design using genetic algorithms. In: First International Conference on Genetic Algorithms in Engineering Systems: Innovations and Applications, pp. 238–244. IET (1995)

27. Pai, N.S., Chang, S.C., Huang, C.T.: Tuning PI/PID controllers for integrating processes with deadtime and inverse response by simple calculations. J. Process Control **20**(6), 726–733 (2010)

28. Piltan, F., Boroomand, B., Jahed, A., Rezaie, H.: Methodology of mathematical error-based tuning sliding mode controller. Int. J. Eng. **6**(2), 96–117 (2012)

29. Proaño, P., Capito, L., Rosales, A., Camacho, O.: A dynamical sliding mode control approach for long deadtime systems. In: 2017 4th International Conference on Control, Decision and Information Technologies (CoDIT), pp. 0108–0113. IEEE (2017)

30. Rajabioun, R.: Cuckoo optimization algorithm. Appl. Soft Comput. **11**(8), 5508–5518 (2011)

31. Rojas, R., Camacho, O., González, L.: A sliding mode control proposal for open-loop unstable processes. ISA Trans. **43**(2), 243–255 (2004)

32. Sira-Ramírez, H.: Dynamical sliding mode control strategies in the regulation of nonlinear chemical processes. Int. J. Control **56**(1), 1–21 (1992)

33. Sira-Ramirez, H., Llanes-Santiago, O.: Dynamical discontinuous feedback strategies in the regulation of nonlinear chemical processes. IEEE Trans. Control Syst. Technol. **2**(1), 11–21 (1994)

34. Utkin, V.: Variable structure systems with sliding modes. IEEE Trans. Autom. Control **22**(2), 212–222 (1977). https://doi.org/10.1109/TAC.1977.1101446

35. Utkin, V., Lee, H.: Chattering problem in sliding mode control systems. In: 2006 International Workshop on Variable Structure Systems, VSS 2006, pp. 346–350. IEEE (2006). https://doi.org/10.1109/VSS.2006.1644542

36. Utkin, Vadim, Poznyak, Alex, Orlov, Yury V.., Polyakov, Andrey: Road Map for Sliding Mode Control Design. SpringerBriefs in Mathematics, Springer, Cham (2020). https://doi.org/10.1007/978-3-030-41709-3

37. Young, K., Utkin, V., Ozguner, U.: A control engineer's guide to sliding mode control. IEEE Trans. Control Syst. Technol. **7**(3), 328–342 (1999). https://doi.org/10.1109/87.761053, http://ieeexplore.ieee.org/document/761053/

38. Yuan, Z., Montes de Oca, M., Birattari, M., Stützle, T.: Continuous optimization algorithms for tuning real and integer parameters of swarm intelligence algorithms. Swarm Intell. **6**, 49–75 (2012). https://doi.org/10.1007/s11721-011-0065-9

Optimizing a Dynamic Sliding Mode Controller with Bio-Inspired Methods: A Comparison

Jorge Espin[1,2] , Sebastian Estrada[1,2] , Diego S. Benítez[2(✉)] ,
and Oscar Camacho[2]

[1] Department of Electronics Engineering, Universidad Técnica Federico Santa María,
Valparaíso, Chile
{jorge.espin,juan.estradac}@usm.cl
[2] Colegio de Ciencias e Ingenierías "El Politécnico",
Universidad San Francisco de Quito USFQ, Quito 170157, Ecuador
{dbenitez,ocamacho}@usfq.edu.ec

Abstract. In the past few years, bio-inspired optimization algorithms
have shown to be an excellent way to solve a wide range of complex
computing problems in science and engineering. This paper compares
bio-inspired algorithms to better understand and measure how well they
find the best tuning parameters for a Dynamic Sliding Mode Control for
integrating systems with an inverse response and dead time. The comparison includes four bioinspired algorithms: particle swarm optimization,
artificial bee colony, ant colony optimization, and genetic algorithms.
It shows how they can improve the performance of the controller by
looking for the best tuning parameter solutions. The parameters of each
algorithm affect the searching mechanism in different ways, and these
effects were tested in two simulated systems. Ant colony optimization is
much better than other algorithms at finding the best answers to our
problems.

Keywords: Bioinspired optimization algorithms · Integrating
systems · Inverse response · Dead time · Dynamical Sliding Mode
Control

1 Introduction

The Sliding Mode Controller (SMC) is a nonlinear control that has been proven
to have the ability to maintain control stability in processes that subject it to
disturbances as well as variations in the system parameters [3, 24]. Sliding mode
control has several benefits, one of which is that it is robust to disturbances and
unmodeled dynamics. Additionally, it is capable of overcoming the uncertainty
of the system. Primarily, it has a positive influence on controlling the nonlinear
system as a whole. On the other hand, the approach is dependent on the state
of the system, and there is a chattering phenomenon in the control signal, both
of which significantly impact the practical use of the method [25].

A. D. Orjuela-Cañón et al. (Eds.): ColCACI 2022, CCIS 1746, pp. 63–80, 2023.
https://doi.org/10.1007/978-3-031-29783-0_5

Several methods can mitigate the chattering problem that can occur with sliding mode control [24,25]. One is the Dynamic Sliding Mode Controller (DSMC), an SMC control strategy with added dynamics to increase system stability and fulfill desired system behavior and performance. This approach was presented in [7,11,22]. Therefore, DSMC is a workable option for decreasing chattering, and it has the bonus of not affecting controller performance owing to the absence of smoothing characteristics. This is a significant advantage. In addition, the convenience of this feature makes it useful for controlling chemical processes, which is important since discontinuities on the final control element are not permitted due to the negative impact they have on product quality.

The performance of SMC or DSMC is associated with tuning parameters; thus, the tuning method of controllers can be used for effectively determining the overall performance [4]. However, most of the time, SMC and DSMC are tuned by the trial and error effort of the designer. Still, it does not offer the best solution, and to reduce this problem, we are discussing and comparing several optimization approaches in this work.

The optimization process aims to discover the best solution(s) to a given problem. Therefore, selecting an appropriate algorithm is critical for achieving this goal [16]. However, some issues are so complicated that finding all feasible solutions is challenging. Thus, several metaheuristic algorithms have been created in the literature to emulate the biological behavior of animal or insect groups by specifying deterministic or random rules to address various optimization problems [5] so that they mimic their natural comportment and reactions in different situations in nature, for instance, as in getting their food. Thus, alternative optimization algorithms to solve complicated problems can be devised. In such a sense, for example, the Artificial Bee Colony (ABC) algorithm [14] simulates the collaborative behavior of bee colonies, and the Particle Swarm Optimization (PSO) algorithm [26] simulates the biological behavior of fish schooling and bird flocking. While the Ant Colony Optimization (ACO) [2] is a class of optimization algorithms modeled on the actions of an ant colony [5]. Other evolutionary algorithms, such as Genetic Algorithms (GA) [19] implement a heuristic search inspired by the natural selection process that mimics biological evolution.

Various bio-inspired optimization algorithms have been successfully used in optimization problems related to mechanical design [13] and control applications for electrical systems [21]. However, as noted in [8], not every algorithm will give the best solutions to every problem. Therefore, this paper compares traditional bio-inspired algorithms to better understand and measure their efficacy in searching for the best tuning parameters for a Dynamical Sliding Mode Control for integrating systems with an inverse response and dead time. The comparative analysis includes four bioinspired algorithms, PSO, ABC, ACO, and GA, and how they can improve the performance of the controller by searching for optimum tuning parameter solutions. The parameters of each algorithm have different impacts on the searching mechanism and are evaluated in two simulated systems.

2   Fundamentals

2.1   Sliding Mode Control Theory

The sliding mode controller is a model-based controller, where the main advantage is that it can compensate for inaccuracies in the process modeling. The SMC contains two terms: the equivalent control part, and the second is known as the reaching mode control law. The first is in charge of keeping the dynamics of the controlled system on a sliding surface, which symbolizes the ideal closed-loop behavior. In order to obtain the desired surface, the second control law is designed. The initial stage in SMC is choosing a sliding surface, commonly expressed as a linear function system state. The proposed sliding equation comprises the reference signal, the model output, and the modeling error.

2.2   Dynamic Sliding Mode Control

The Dynamic Sliding Mode Controller (DSMC) is a control method that incorporates additional dynamics into the SMC. The additional dynamics are intended to achieve and improve system stability and accomplish desirable system behavior and performance. As a result, DSMCs are a viable solution for reducing the nocive chattering effects original in the SMC. Because no smoothing functions are used, they have the advantage of not lowering controller performance. This feature is suited for chemical process control where discontinuities on the final element of control are not permitted and quick vibrations are not tolerated due to undesirable effects on product quality [11].

2.3   Particle Swarm Optimization

Particle Swarm Optimization (PSO) is a biologically inspired computer search and optimization approach created by Eberhart and Kennedy in 1995. It is based on the social behaviors of flocking birds or schooling fish. In order to increase the speed of convergence and the quality of the solution found by the PSO, several basic variants have been created. Basic PSO, on the other hand, is better suited to solving static, simple optimization problems [26].

2.4   Artificial Bee Colony

The artificial bee colony (ABC) algorithm is a tool for simulating honey bee foraging behavior. Karaboga proposed ABC in 2005 [14] as part of a group of swarm intelligence algorithms. Each candidate solution in this method is the position of the food source in the search space, and the fitness evaluator is the quality of the nectar amount of the food source. Employed bees, bystanders, and scouts are the three types of bees involved. The number of worker bees is equal to the number of food sources. Employed bees depart the hive in quest of a food source, collecting nectar from different food sources in the region of the one they find. When they return to their hive, they execute a dance in which they alert bystanders about the newly discovered food supply (location and quality). Onlookers use the information offered by the employed bees to choose a new food source based on the selection. This process is repeated until the optimal food source is found [23].

2.5   Ant Colony Optimization

Some foraging behavior of ant species inspires ant colony optimization (ACO). These ants leave pheromones on the ground to identify a suitable path for other colony members to follow. For solving optimization problems, ant colony optimization uses a similar method. Combining a priori knowledge about the structure of a potential solution with posterior information about the structure of previously obtained reasonable solutions is a crucial feature of ACO algorithms.

2.6   Genetic Algorithms

The genetic algorithm (GA) is a model or abstraction of biological evolution based on Charles Darwin's theory of natural selection, created by John Holland and his coworkers in the 1960s and 1970s. In studying adaptive and artificial systems, Holland was arguably the first to use crossover and recombination, mutation, and selection [19]. The genetic algorithm as a problem-solving approach is incomplete without these genetic operators. Genetic algorithms have several advantages over classic optimization algorithms. The ability to deal with complex problems and parallelism is the most significant. Furthermore, genetic algorithms can handle anything, whether the objective (fitness) function is stable or non-stationary (changes with time), linear or nonlinear, continuous or discontinuous, or random noise.

3   DSMC Parameters Tuning Using Bio-Inspired Optimization Techniques

The Dynamic Sliding Mode Controller for Inverse Response Integrating Systems (DSMC-IRIS) with high delay proposed by Espin and Camacho in [6], is a control strategy based on sliding mode theory and internal model. In addition, it incorporates a Smith predictor to compensate for the long-dead time and an Iinoya and Alpeter approach to mitigate the inverse condition, as shown in Fig. 1.

A PID sliding surface is used for the controller design, as presented in (1).

$$S(t) = \lambda_1 e(t) + \lambda_0 \int_0^t e(t)\, dt + \frac{de(t)}{dt} \tag{1}$$

The process characterization corresponds to a reduced-order inverse-response integrating model with dead time, as detailed in (2).

$$G_p(s) = K \frac{-\eta s + 1}{s(\tau s + 1)} e^{-t_0 s} \tag{2}$$

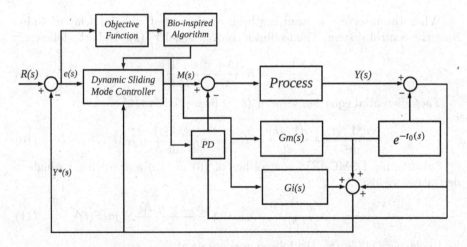

Fig. 1. DSMC-IRIS control scheme proposal.

The full expression of the model-based DSMC-IRIS [6] is obtained in (3).

$$\dot{M}(t) = \frac{1}{K(\lambda - \eta)}[(1 - \lambda_1)\dot{Y}^*(t) + \tau\lambda_0 e(t) - KM(t)]...$$

$$... + K_d Sign(S(t)) \quad (3)$$

A classical PD is part of the hybrid control topology of Fig. 1, defined by:

$$PD = K_o(T_d s + 1) \quad (4)$$

$G_m(s)$ corresponds to the Smith predictor fast model,

$$G_m(s) = K\frac{-\eta s + 1}{s(\tau s + 1)} \quad (5)$$

and $G_i(s)$ is related to the Iinoya and Alpeter approach [12], responsible for mitigating the inverse condition,

$$G_i(s) = K\frac{\lambda s}{s(\tau s + 1)} \quad (6)$$

3.1 Lyapunov Stability Condition

The Lyapunov stability condition concerns the inequality of (7):

$$\dot{V} = S\dot{S} < 0 \quad (7)$$

$\dot{S}(t)$ is detailed as follows:

$$\frac{dS(t)}{dt} = -\frac{d^2Y^*(t)}{dt^2} - \lambda_1\frac{dY^*(t)}{dt} + \lambda_0 e(t) \quad (8)$$

When the inverse condition has been compensated and the delay excluded from the control system. The feedback control signal $Y^*(s)$ yields the following:

$$\frac{Y^*(s)}{M(s)} = K\frac{[(\lambda - \eta)s + 1]}{s(\tau s + 1)} \tag{9}$$

The differential equation form of (9) is presented in (10)

$$\frac{\tau d^2 Y^*(t)}{dt^2} + \frac{dY^*(t)}{dt} = K(\lambda - \eta)\frac{dM(t)}{dt} + KM(t) \tag{10}$$

Substituting DSMC-IRIS control law of (3) in (10) and solving its highest derivative, yields:

$$\frac{d^2 Y^*(t)}{dt} = -\lambda_1\frac{dY^*(t)}{dt} + \lambda_0 e(t) + \frac{K(\lambda - \eta)K_d}{\tau}Sign(S(t)) \tag{11}$$

Replacing (11) in (8), the following is obtained:

$$\frac{dS(t)}{dt} = -\frac{K(\lambda - \eta)K_d}{\tau}Sign(S(t)) \tag{12}$$

Therefore, the reaching condition is defined by:

$$S\dot{S} = -\frac{K(\lambda - \eta)K_d}{\tau}S(t)Sign(S(t)) < 0 \tag{13}$$

$$-\frac{K(\lambda - \eta)K_d}{\tau}|S(t)| < 0$$

where,

$$\mathbb{K} = \frac{K(\lambda - \eta)K_d}{\tau} \tag{14}$$

So,

$$-\mathbb{K}|S(t)| < 0 \tag{15}$$

In order to guarantee the controller stability, it is mandatory to make sure that $\mathbb{K} > 0$.

3.2 Optimal Tuning Parameters from the Application of the Different Bio-Inspired Methods

As reviewed in the previous section, the parameters of interest are λ_1, λ_0, K_d, K_o, T_d, and λ. The large number of tuning parameters suggests using optimization methods to obtain optimal solutions. However, in [6] the trial and error technique was used.

The parameter solutions were found using different optimization algorithms, such as ABC, PSO, ACO, and GA. In all cases, the objective function J to be minimized is given by the Integral of Squared Error (ISE), as shown in (16).

$$J = ISE = \int_0^t e^2(t)dt \tag{16}$$

The bio-inspired optimization methods were executed in a 3.1 GHz Dual-Core Intel Core i7 PC with 16 GB of RAM and a macOS operating system. While the simulation was developed in Matlab, the source codes for each of the optimization algorithms were obtained from PSO [9], ACO [10], GA [17], and ABC [1], respectively.

4   Experimentation and Simulation Results

Two linear inverse-response integrating systems with dead time and particular characteristics were considered to compare the aforementioned bio-inspired optimization techniques for DSMC-IRIS tuning. The quantitative analysis is based on performance indexes such as Integral of Squared Error (ISE), Total Variation of control effort (TVu), transient characteristics (Maximum Overshoot (OS%), and Settling Time (Ts)). The Execution Time (ET) was also considered to verify the fastest implemented algorithm.

4.1   High-Order Linear Integrating System with Inverse Response and Dead Time

Example 1: The integrating high-order linear inverse-response system with delay expressed in (17), and studied by Pai et al. in [20].

$$G_p(s) = \frac{0.5(-0.5s + 1)}{s(0.5s + 1)(0.4s + 1)(0.1s + 1)} e^{-0.7s} \tag{17}$$

The DSMC-IRIS design requires an integrating inverse response reduced-order model with delay. Hence, Luyben's computational identification method [18] was used. Yielding,

$$G_{pm}(s) = \frac{0.583(-0.4699s + 1)}{s(1.1609s + 1)} e^{-0.81s}$$

All the bio-inspired optimization methods used a population size, $N_p = 20$, and iterations, $T = 5$. The PSO velocities C1 and C2 are equal to 2. GA used a % of crossover of 0.8 and % of mutation of 0.2, and the tournament selection method was chosen. ACO used an Intensification Factor of 0.5 and a Deviation-Distance Ratio of 1. Finally, ABC considered the number of the onlooker and employed bees to equal half the colony size, respectively.

After numerous simulations, the Fig. 2 shows the optimal solutions for DSMC-IRIS parameter tuning of Example 1 (Tables 1, 2, and 3).

Table 1. Execution time of bio-inspired algorithms for Example 1

	IRIS	ABC	ACO	GA	PSO
ET	NA	968.80 s	48.46 s	289.51 s	236.29 s

Table 2. Tuning parameters for Example 1

Controller	λ_1	λ_0	K_d	K_o	T_d	λ
DSMC-IRIS	2.096	1.098	1.294	0.709	0.788	3.000
DSMC-IRIS-GA	1.251	1.602	1.978	0.921	0.353	3.920
DSMC-IRIS-PSO	2.079	2.782	1.000	0.704	0.790	3.242
DSMC-IRIS-ABC	1.308	1.516	1.929	0.800	0.583	3.112
DSMC-IRIS-ACO	1.475	1.466	1.262	0.626	0.709	3.387

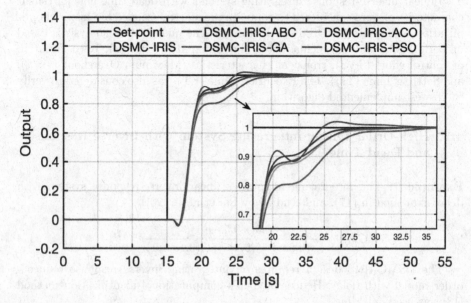

Fig. 2. Comparison of different bio-inspired optimization methods for DSMC-IRIS tuning of Example 1 (tracking test).

Table 3. Performance indexes and transient characteristics of Example 1 (tracking test)

Controller	ISE	TVu	Ts [s]	OS [%]
DSMC-IRIS	0.770	27.704	26.330	2.080
DSMC-IRIS-GA	0.592	17.521	28.127	0.000
DSMC-IRIS-PSO	0.699	8.220	26.300	0.000
DSMC-IRIS-ABC	0.643	13.600	25.252	0.000
DSMC-IRIS-ACO	0.714	10.826	26.421	0.009

Figure 2 depicts the corresponding responses for a step change at t = 20 s. Comparing the applied bio-inspired algorithms, we identify some important issues to mention. DSMC-IRIS-GA has the slowest response with Ts = 28.127 s and is caused by having the highest value of $\lambda = 3.920$. By simulation and prior knowledge, this parameter is commonly associated with generating a smoother

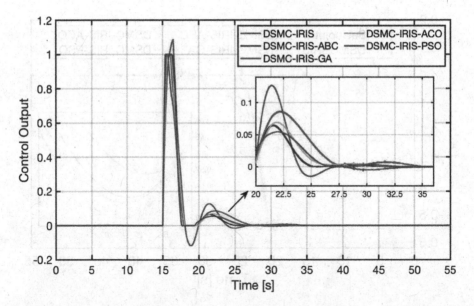

Fig. 3. Comparison of the outputs for the controllers of Example 1 (tracking test).

response but with a long settling time. DSMC-IRIS-ACO and DSMC-IRIS-PSO have similar characteristics. However, the former outperforms the latter in ET, which is a tremendous advantage in computational processing. Regarding DSMC-IRIS-ABC, this approach provides a lower Ts. However, the main problem of this algorithm is related to the ET, which is equal to 968.80 s, implying that the applied ABC algorithm is at least 20 times slower than the applied ACO algorithm. All mentioned cases above have particularities to analyze; nevertheless, all applied bio-inspired algorithms give a better controller performance than the previous DSMC-IRIS, which was tuned by a trial-and-error technique (Table 4 and Fig. 3).

Table 4. Performance indexes for load disturbance test of Example 1

Controller	ISE	TVu
DSMC-IRIS	0.224	27.605
DSMC-IRIS-GA	1	42.581
DSMC-IRIS-PSO	0.294	21.191
DSMC-IRIS-ABC	0.426	41.477
DSMC-IRIS-ACO	0.289	26.812

4.2 Linear Integrating Inverse Response System with Long Dead Time and Time Constant

Example 2: The linear integrating inverse response system with long-dead time and time constant shown in (18) and considered by Kaya in [15] (Figs. 4 and 5).

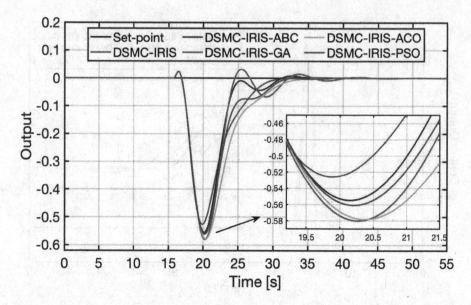

Fig. 4. Comparison of different bio-inspired optimization methods for DSMC-IRIS tuning of Example 1 (load disturbance test).

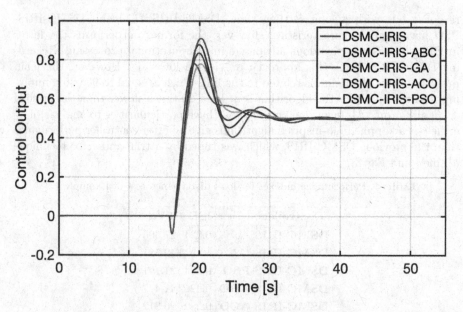

Fig. 5. Comparison of the outputs for the controllers of Example 1 (load disturbance test).

$$G_p(s) = \frac{(-s+1)}{s(10s+1)}e^{-7s} \qquad (18)$$

For all the bio-inspired optimization methods, the population size was set to $N_p = 15$, and the number of iterations was set to $T = 20$. The C1 and C2 velocities for the PSO were equal to 2. The GA used a % crossover of 0.8 and % of mutation of 0.2, and the tournament selection method was chosen. ACO used an Intensification Factor of 0.5 and a Deviation-Distance Ratio of 1. Finally, ABC considered the number of the onlooker and employed bees to equal half the colony size, respectively (Tables 5, 6, 7, 8 and Figs. 6, 7).

Table 5. Execution time of bio-inspired algorithms for Example 2

	IRIS	ABC	ACO	GA	PSO
ET	NA	3820.14 s	135.22 s	237.61 s	91.33 s

Table 6. Tuning parameters for Example 2

Controller	λ_1	λ_0	K_d	K_o	T_d	λ
DSMC-IRIS	0.243	0.015	5.608	0.080	7.000	15
DSMC-IRIS-GA	2.085	0.069	10.273	0.075	9.919	9.222
DSMC-IRIS-PSO	1.613	0.054	2.221	0.072	7.642	9.435
DSMC-IRIS-ABC	0.436	0.066	7.454	0.079	6.568	8.951
DSMC-IRIS-ACO	2.234	0.200	11.068	0.055	9.825	9.419

Table 7. Performance indexes and transient characteristics of Example 2 for tracking test

Controller	ISE	TVu	Ts [s]	OS [%]
DSMC-IRIS	2.187	50.236	59.700	0.663
DSMC-IRIS-GA	0.358	771.266	52.250	0.276
DSMC-IRIS-PSO	0.700	166.655	51.601	0.575
DSMC-IRIS-ABC	1.183	559.301	45.379	0.000
DSMC-IRIS-ACO	0.338	831.084	53.176	0.006

Table 8. Performance indexes for load disturbance test of Example 2

Controller	ISE	TVu
DSMC-IRIS	1	47.139
DSMC-IRIS-GA	0.044	730.311
DSMC-IRIS-PSO	0.024	157.066
DSMC-IRIS-ABC	0.575	526.372
DSMC-IRIS-ACO	0.018	786.739

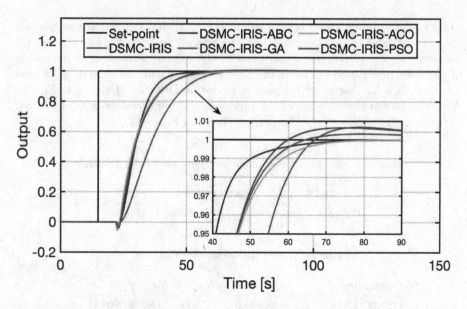

Fig. 6. Comparison of different bio-inspired optimization methods for DSMC-IRIS tuning of Example 2 (tracking test).

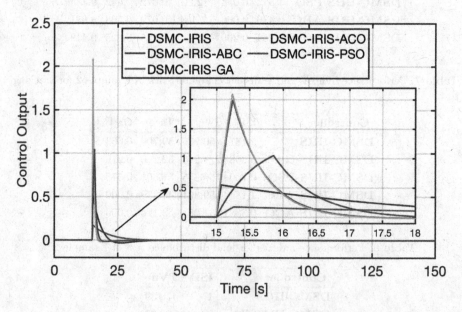

Fig. 7. Comparison of the outputs for the controllers of Example 2 (tracking test).

Figure 8 shows the responses of applying a step change and a disturbance of -0.1 at t = 250 s. All the applied bio-inspired algorithms have some essential features to analyze. Referring to the λ term, as in Example 1, the controller

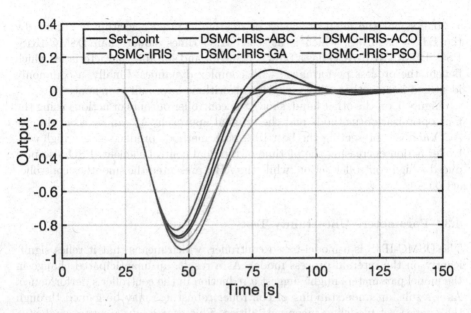

Fig. 8. Comparison of different bio-inspired optimization methods for DSMC-IRIS tuning of Example 2 (load disturbance test).

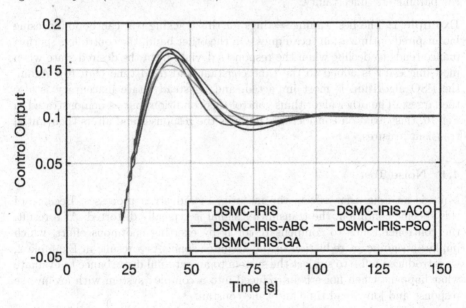

Fig. 9. Comparison of the outputs for the controllers of Example 2 (load disturbance test).

with the highest value DSMC-IRIS has the slowest response reflected in Ts = 44.67 s. On the other hand, DSMC-IRIS-ABC is the fastest one with a Ts

= 30.38 s, but the main drawback of this bio-inspired algorithm is verified by
the ET = 3820.14 s, which is at least 41.83 times slower than DSMC-IRIS-
PSO. DSMC-IRIS-ACO has a typical behavior and suitable characteristics which
benefit the process performance with complex dynamics. Finally, a commonly
identified issue is that all bio-inspired algorithms have null overshoot.

Figure 9, on the other hand, shows the controller outputs or actions using the
four optimization methods and the original approach. As can be seen, despite
ACO and GA presenting the best ISE, both methods produce a very high controller action decreasing the lifetime of the final control element. PSO also produced a high controller action, while the ABC presented the smoothes controller
action.

4.3   Parameters Uncertainty Test

The DSMC-IRIS is a model-based controller, which means that it relies significantly on the internal process model. As a result, an unanticipated change in
the model parameters might suggest a reduction in the controller's performance.
As a result, an understanding of controller robustness may be gained through
the analysis of modeling incompatibilities. This case study aims to conduct a
comparative analysis of several bio-inspired optimization approaches to assess
the parameters' uncertainty.

Example 1: The best tuning settings for this tracking test can be found using
bio-inspired optimization techniques. On the other hand, the controllers' performance tends to decline when the responses deviate from the desired state when
modeling error is added to the time constant and dead time $(tau, t0)$. Again,
the PSO algorithm is most impacted, and its steady-state inaccuracy is also
the largest. The other algorithms control this condition, but as demonstrated in
Fig. 10, this variation considerably impacts the responses and alters their initial
transient features.

4.4   Noise Test

Noise is an issue that affects the majority of industrial processes. Because of
this specific condition, the transmitter signal is typically distorted. As a result,
the controller tends to compensate for it by exerting enormous effort, which
implies a significant reduction in the actuator span. As a result, in Example 2,
we introduced noise to subject the system to an external disturbance to evaluate
what happens when noise is introduced into a complex system with an inverse
response and long dead time and time constant.

Example 2: The process response to noise is depicted in Figs. 12 and 13. Each
controller can achieve the target value, and their transient characteristics are
remarkably comparable to those without noise. The most astonishing finding
is depicted in Fig. 13, where the noise effect is discernible. However, none of
the tuning strategies based on bio-inspired optimization are influenced by noise
and retain a smooth control effort. This graph illustrates how DSMC-IRIS may

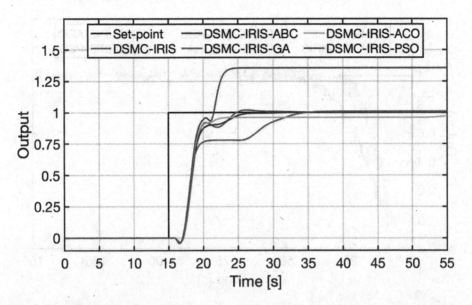

Fig. 10. Comparison of the response of the controllers of Example 1 for parameters uncertainty test.

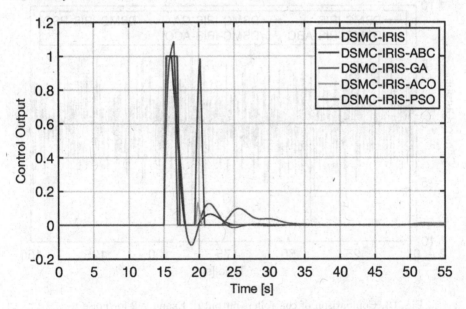

Fig. 11. Comparison of controllers output of Example 1 for parameters uncertainty test.

impact the actuator's usable lifetime if implemented in a real-world situation. In contrast, the other controllers are sturdy enough to withstand loud industrial situations (Fig. 11).

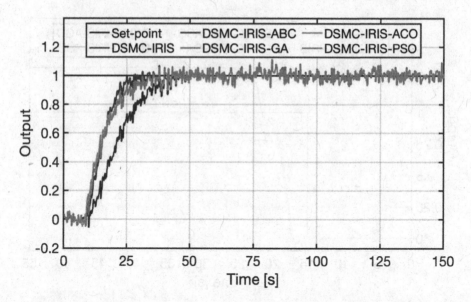

Fig. 12. Comparison of the response of the controllers of Example 2 for noise test.

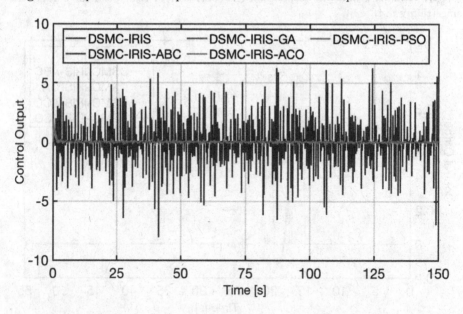

Fig. 13. Comparison of controllers output of Example 2 for noise test.

5   Conclusion

In this work, bio-inspired algorithms were investigated to find the most appropriate tuning parameters for a Dynamic Sliding Mode Control, which integrates systems characterized by an inverse response and a dead time. The comparative study examined how four bioinspired algorithms, PSO, ABC, ACO, and GA, may improve the controller's performance by looking for the optimal tuning parameter solutions. When simulated in two different integrating systems with an inverse response and dead time, certain performance indices of each method revealed distinct effects on the searching mechanism. These results were obtained through the evaluation. According to the findings, all the algorithms could locate a solution; however, the ACO demonstrated a competence much superior to other algorithms when locating optimum solutions considering execution time.

Acknowledgment. J. Espin and S. Estrada would like to express their gratitude to the Advanced Control Systems Research Group at USFQ for the opportunity to participate in the research internship.

References

1. Artificial Bee Colony (ABC) Algorithm. Intelligent Systems Research Group Department of Computer Engineering at Erciyes University (2009). https://abc.erciyes.edu.tr/
2. Al Salami, N.M.: Ant colony optimization algorithm. UbiCC J. **4**(3), 823–826 (2009)
3. Báez, E., Bravo, Y., Leica, P., Chávez, D., Camacho, O.: Dynamical sliding mode control for nonlinear systems with variable delay. In: 2017 IEEE 3rd Colombian Conference on Automatic Control (CCAC), pp. 1–6. IEEE (2017)
4. Camacho, O., Smith, C.A.: Sliding mode control: an approach to regulate nonlinear chemical processes. ISA Trans. **39**(2), 205–218 (2000)
5. Darwish, A.: Bio-inspired computing: algorithms review, deep analysis, and the scope of applications. Future Comput. Inform. J. **3**(2), 231–246 (2018)
6. Espín, J., Camacho, O.: A proposal of dynamic sliding mode controller for integrating processes with inverse response and deadtime. In: 2021 IEEE Fifth Ecuador Technical Chapters Meeting (ETCM), pp. 1–6. IEEE (2021)
7. Espín, J., Castrillon, F., Leiva, H., Camacho, O.: A modified smith predictor based-sliding mode control approach for integrating processes with dead time. Alex. Eng. J. **61**(12), 10119–10137 (2022)
8. Game, P.S., Vaze, D.V., M, D.E.: Bio-inspired optimization: metaheuristic algorithms for optimization (2020). https://doi.org/10.48550/ARXIV.2003.11637,https://arxiv.org/abs/2003.11637
9. Gopal, A., Sultani, M.M., Bansal, J.C.: On stability analysis of particle swarm optimization algorithm. Arab. J. Sci. Eng. **45**(4), 2385–2394 (2020). https://doi.org/10.1007/s13369-019-03991-8
10. Heris, M.K.: Ant colony optimization for continuous domains (ACOR) (2015). https://www.mathworks.com/matlabcentral/fileexchange/52860-ant-colony-optimization-for-continuous-domains-acor. MATLAB Central File Exchange. Accessed 7 Mar 2022
11. Herrera, M., Camacho, O., Leiva, H., Smith, C.: An approach of dynamic sliding mode control for chemical processes. J. Process Control **85**, 112–120 (2020)
12. Iinoya, K., Altpeter, R.J.: Inverse response in process control. Ind. Eng. Chem. **54**(7), 39–43 (1962)

13. Johnvictor, A.C., Durgamahanthi, V., Pariti Venkata, R.M., Jethi, N.: Critical review of bio-inspired optimization techniques. Wiley Interdisc. Rev.: Comput. Stat. **14**(1), e1528 (2022)

14. Karaboga, D., et al.: An idea based on honey bee swarm for numerical optimization. Tech. rep., Technical report-tr06, Erciyes University, Engineering Faculty, Computer Engineering Department, Kayseri/Türkiye (2005)

15. Kaya, I.: Controller design for integrating processes with inverse response and dead time based on standard forms. Electr. Eng. **100**(3), 2011–2022 (2018)

16. Korošec, P., Melab, N., Talbi, E.G.: Bioinspired Optimization Methods and Their Applications: 8th International Conference, BIOMA 2018, Paris, France, May 16–18, 2018, Proceedings, vol. 10835. Springer, Cham (2018). https://doi.org/10.1007/978-3-319-91641-5

17. Labs, S.: Genetic algorithm (2020) https://www.mathworks.com/matlabcentral/fileexchange/74132-genetic-algorithm. MATLAB Central File Exchange. Accessed 7 Mar 2022

18. Luyben, W.L.: Identification and tuning of integrating processes with deadtime and inverse response. Ind. Eng. Chem. Res. **42**(13), 3030–3035 (2003)

19. Mirjalili, S.: Genetic algorithm. In: Evolutionary Algorithms and Neural Networks. SCI, vol. 780, pp. 43–55. Springer, Cham (2019). https://doi.org/10.1007/978-3-319-93025-1_4

20. Pai, N.S., Chang, S.C., Huang, C.T.: Tuning PI/PID controllers for integrating processes with deadtime and inverse response by simple calculations. J. Process Control **20**(6), 726–733 (2010)

21. Roni, M.H.K., Rana, M.S., Pota, H.R., Hasan, M.M., Hussain, M.S.: Recent trends in bio-inspired meta-heuristic optimization techniques in control applications for electrical systems: a review. Int. J. Dyn. Control 1–13 (2021). https://doi.org/10.1007/s40435-021-00892-3

22. Sira-Ramírez, H.: Dynamical sliding mode control strategies in the regulation of nonlinear chemical processes. Int. J. Control **56**(1), 1–21 (1992)

23. Ting, T.O., Yang, X.-S., Cheng, S., Huang, K.: Hybrid metaheuristic algorithms: past, present, and future. In: Yang, X.-S. (ed.) Recent Advances in Swarm Intelligence and Evolutionary Computation. SCI, vol. 585, pp. 71–83. Springer, Cham (2015). https://doi.org/10.1007/978-3-319-13826-8_4

24. Utkin, V., Poznyak, A., Orlov, Y., Polyakov, A.: Conventional and high order sliding mode control. J. Franklin Inst. **357**(15), 10244–10261 (2020)

25. Utkin, V., Poznyak, A., Orlov, Y.V., Polyakov, A.: Road Map for Sliding Mode Control Design. Springer, Cham (2020). https://doi.org/10.1007/978-3-030-41709-3

26. Wang, D., Tan, D., Liu, L.: Particle swarm optimization algorithm: an overview. Soft. Comput. **22**(2), 387–408 (2018). https://doi.org/10.1007/s00500-016-2474-6

A Robust Controller Based on LAMDA and Smith Predictor Applied to a System with Dominant Time Delay

Luis Morales[1]([✉]) [iD], Oscar Camacho[2] [iD], and Paulo Leica[1] [iD]

[1] Departamento de Automatización y Control Industrial, Escuela Politécnica Nacional, Quito 170525, Ecuador
{luis.moralesec,paulo.leica}@epn.edu.ec

[2] Colegio de Ciencias e Ingenierías "El Politécnico", Universidad San Francisco de Quito USFQ, Quito 170157, Ecuador
ocamacho@usfq.edu.ec

Abstract. This document presents a LAMDA (Learning Algorithm for Multivariable Data Analysis) Sliding-Mode Control LSMC applied to a pH neutralization reactor with dominant dead time. Due to the non-linearity generated by the dead time, the application of an additional Smith Predictor structure is proposed to improve the system's response when reference changes and disturbances occur around the operating point in which the system works. The controller is validated through different simulations in which it is evident that the proposed approach is stable in controlling the pH neutralization reactor.

Keywords: LAMDA · Smith Predictor · Dominant Dead Time · Sliding-Mode Control

1 Introduction

SISO uncertain systems are a study area of great interest in control systems research. Most processes have nonlinear characteristics that make it challenging to obtain exact models that facilitate the design of controllers, especially when the operating point of the process to be controlled changes the intrinsic characteristics of the model [1, 2]. Classic controllers such as the PID are still the most used for industrial applications [3]. It is a controller that depends on fine-tuning its scaling gains, established based on an operating system point. However, if the process operating point changes, the controller deteriorates its performance, becomes unstable, and requires a new calibration.

Intelligent controllers based on Fuzzy Logic (FL) present excellent performance in systems in which the model of the system to be controlled is not exactly known [4]. In addition, FL controllers present excellent efficiency against uncertainty [5]; through FL, nonlinear controllers are designed empirically without mathematical complications.

On the other hand, Sliding-Mode Control (SMC) is a robust method used explicitly to control systems with model uncertainty and is insensitive to external disturbances [6].

© The Author(s), under exclusive license to Springer Nature Switzerland AG 2023
A. D. Orjuela-Cañón et al. (Eds.): ColCACI 2022, CCIS 1746, pp. 81–98, 2023.
https://doi.org/10.1007/978-3-031-29783-0_6

The drawback of the SMC is the phenomenon known as chattering, which is a high-frequency oscillation present in the output control action that, in real applications, can damage actuators (mechanical parts), excite unmodeled high-frequency dynamics, and degrade the entire control system, which can cause unpredictable instability [7].

In the FL and robust controllers field, we have proposed a controller that bases its operation on the LAMDA (Learning Algorithm for Multivariate Data Analysis) algorithm, an artificial intelligence method for classification and clustering tasks.

LAMDA initially computes the Marginal Adequacy Degree (MAD), a parameter that measures the contribution of the descriptors of an object to each cluster/class, with fuzzy probability functions like the Gaussian function. Then, using fuzzy aggregation operators, the MADs in each class are combined to obtain the Global Adequacy Degree (GAD), a parameter that quantifies the membership degree of any individual to each class of the system. Finally, LAMDA identifies and assigns the individual (object) to the most suitable class where the maximum GAD is computed.

In [4], we formalize the LAMDA algorithm for control based on the Lyapunov and SMC theory fundamentals to guarantee stability and robustness. The controller is called LAMDA-SMC (LSMC) and takes the features of LAMDA to design a chattering-free controller. LSMC has been tested to control chemical processes and robotics [8]. Its operation has been validated, showing that it can control systems with high non-linearity and model uncertainties.

This work presents the design of the LSMC controller applied to a pH neutralization reactor with dominant dead time and variable modeling parameters [9]. Furthermore, the use of a control structure based on a Smith Predictor is proposed to improve the response of the system when it is subjected to reference changes (and therefore the change of the operating point) and disturbances, for which qualitative and quantitative analysis is made based on the measurement of performance indices.

The paper is organized as follows: Sect. 2 presents the fundamentals of LAMDA-SMC and the Smith Predictor. Section 3 applied the LSMC in a Smith Predictor scheme (LSMC-SP) to a pH neutralization reactor. Section 4 presents the simulations of the proposed controller, evaluating its performance with other controllers such as PID, SMC, and LSMC to make a comparative analysis. Finally, conclusions are presented in Sect. 5.

2 Background

2.1 LAMDA as Controller

LAMDA is a fuzzy algorithm for classification/clustering that computes a similitude degree between the descriptors that characterize an object $O = [o_1; \ldots; o_j; \ldots; o_l] \in \mathfrak{R}$, and the "$m$" classes $C = \{C_1; \ldots; C_k; \ldots; C_m\}$, to identify the class to which the object belongs.

The descriptors must be normalized $\overline{o}_j \in [0, 1]$ as follows:

$$\overline{o}_j = \frac{o_j - o_{jmin}}{o_{jmax} - o_{jmin}} \tag{1}$$

where o_{jmax} is the maximum and o_{jmin} is the minimum value of the descriptor o_j.

Then, it is computed the Marginal Adequacy Degree (MAD). This parameter evaluates the similitude of each object descriptor with the equivalent descriptor in a class k. For the MAD computation, probability density functions like the Gaussian are used. The Gaussian function requires the average of the descriptor j that belongs to the class k ($\rho_{k,j}$):

$$MAD_{k,j} = e^{-\frac{1}{2}\left(\frac{\overline{o}_j - \rho_{k,j}}{\sigma_{k,j}}\right)^2} \tag{2}$$

$$\rho_{k,j} = \frac{1}{n_{k,j}} \sum_{t=1}^{n_{k,j}} \overline{o}_j(t) \tag{3}$$

where $n_{k,j}$ is the number of elements and $\sigma_{k,j}$ is the standard deviation of the descriptor j in the class k, respectively. The parameter $\sigma_{k,j}$ is calculated as:

$$\sigma_{k,j}^2 = \frac{1}{n_{k,j} - 1} \sum_{t=1}^{n_{k,j}} (\overline{o}_j(t) - \rho_{k,j})^2 \tag{4}$$

The MADs are mixed with fuzzy logic operators to compute the Global Adequacy Degree (GAD). The GAD is related to the membership degree of the object O to each class k. The GADs are the product of linear interpolations of the S-norm "$S(a, b)$" and the T-norm "$T(a, b)$" as the Dombi operator [10] defined as:

$$S(a, b) = 1 - \frac{1}{1 + \sqrt[p]{\left(\frac{a}{1-a}\right)^p + \left(\frac{b}{1-b}\right)^p}} \tag{5}$$

$$T(a, b) = \frac{1}{1 + \sqrt[p]{\left(\frac{1-a}{a}\right)^p + \left(\frac{1-b}{b}\right)^p}} \tag{6}$$

where $p \geq 1$ calibrates the sensitivity and a, b are the MADs of the class k determined with the Dombi function [4].

The GAD of the normalized object $\overline{O}$ in each class, k is computed as:

$$GAD_{k,\overline{o}}(MAD_{k,1}, \dots MAD_{k,m}) = \delta T(MAD_{k,1}, \dots, MAD_{k,m})$$
$$+ (1 - \delta)S(MAD_{k,1}, \dots, MAD_{k,m}) \tag{7}$$

where $\delta \in [0, 1]$ is the exigency. Values close to 1 make a stricter algorithm, and values close to 0 make a permissive algorithm.

LAMDA identifies the system's current state in the control field to move it to the desired state using the GADs. The required state has the error and its derivatives equal to zero; for this purpose, we need to define rules based on the system knowledge as in FL control. The linguistic expression that represents the FL inference using the classes of LAMDA is:

$$Rule^{(k)} : IF\ \overline{o}_1\ is\ F_1^p\ and\ \dots \overline{o}_j\ is\ F_j^q\ \dots\ and\ \overline{o}_l\ is\ F_l^r\ THEN\ y_k\ is\ \gamma_k \tag{8}$$

where $\overline{o}_j$ is the object descriptor taking values in the universe of discourse U_j. y_k is in the universe of discourse V. The fuzzy set $F_j = \left\{ F_j^q : q = 1, 2, \ldots, Q \right\}$ belongs to U_j, the fuzzy set γ_k belongs to V, and $Rule^{(k)}$ is the rule applied to the class k.

The LAMDA inference mechanism proposed uses the GADs of each class and the first-order Takagi-Sugeno inference [11]. In (8), γ_k is a singleton value specified for the class k. Thus, the control output for LAMDA is computed as:

$$u = \Gamma \sum_{k=1}^{m} \gamma_k GAD_{k,\overline{O}} \tag{9}$$

$$\Gamma = \left| \frac{\operatorname{argmax}(\gamma_k)}{\sum_{k=1}^{m} \gamma_k GAD_{k,\operatorname{argmax}(\overline{O})}} \right| \tag{10}$$

where u is the output of the LAMDA controller, Γ is an adjustment parameter used as a saturator and γ_k is the weight assigned to the class k.

2.2 LAMDA Sliding-Mode Control (LSMC)

LSMC considers a SISO nonlinear system, defined in state-space as:

$$\dot{x}_i(t) = x_{i+1}(t), i = 1, \ldots, n-1$$

$$\dot{x}_n(t) = A(X(t), t) + b(X(t), t)u(t) + d(t) \tag{11}$$

With: $A(X(t), t)$ and $b(X(t), t)$ nonlinear functions, $X(t)$ is the state vector of the system, $u(t) \in \mathbb{R}$ is the control input, and $d(t) \in \mathbb{R}$ is an unknown disturbance.

The control action is obtained using the following assumptions: The states of the system $X(t)$ are measurable and bounded, $|A(X(t), t)| \leq \beta_A, b(X(t), t) \neq 0$ and $|d(t)| \leq \beta_d$, with β_A, β_d unknown positive factors [11].

The desired state is $X_d(t)$, thus, the tracking error is:

$$E(t) = X_d(t) - X(t) \tag{12}$$

The tracking error to be tracked by the controller should satisfy:

$$\lim_{t \to \infty} \|E(t)\| = \lim_{t \to \infty} \|X_d(t) - X(t)\| \to 0 \tag{13}$$

The sliding surface selected for LSMC is defined as:

$$s(t) = \left(\frac{d}{dt} + \lambda \right)^n \int e(t)dt \tag{14}$$

where: n is the system order, and λ is a strictly positive parameter to define the sliding hyperplane.

If (13) satisfies, then (14) reaches a constant value. To keep $s(t)$ constant, the condition of (15) must be met:

$$\dot{s}(t) = 0 \tag{15}$$

In the SMC proposed in [12], the control law is:

$$u = u_c + u_d \tag{16}$$

where u_c is the continuous control action required to keep the system on the sliding surface and u_d is the discontinuous control action required to reach the sliding surface.

Deriving (14):

$$\dot{s}(t) = e^{(n)}(t) + r_{n-1}\lambda e^{(n-1)}(t) + r_{n-2}\lambda^2 e^{(n-2)}(t) + \cdots + r_1\lambda^{(n-1)}\dot{e}(t) + r_0\lambda^n e(t)$$

$$= e^{(n)}(t) + \sum_{i=1}^{n} r_{n-i}\lambda^i e^{(n-i)}(t) \tag{17}$$

With $\{r_{n-1}, r_{n-2}, \ldots, r_1, r_0\}$ the coefficients obtained by solving the polynomial of (14) with power n, and $e^{(0)}(t) = e(t)$, $r_0 = 1$.

Replacing (11) and (12) in (17):

$$\dot{s}(t) = \dot{x}_{dn}(t) - A(X(t), t) - b(X(t), t)u - d(t) + \sum_{i=1}^{n} r_{n-i}\lambda^i e^{(n-i)} \tag{18}$$

where $\dot{x}_{dn}(t)$ is the n-derivative of the desired reference.

For the definition of the continuous control law, $u = u_{nc}$ in (18):

$$\dot{s}(t) = \dot{x}_{dn}(t) - A(X(t), t) - b(X(t), t)u_{nc} - d(t) + \sum_{i=1}^{n} r_{n-i}\lambda^i e^{(n-i)} \tag{19}$$

Besides, the control action in LAMDA [13] considering $O = [\dot{s}(t)]$ is:

$$u_{nc} = LAMDA_c(\dot{s})$$

$$= \left| \frac{\text{argmax}(\gamma_k)}{\sum_{k=1}^{m} \gamma_k GAD_{k,\text{argmax}(\dot{s}(t))}} \right| \sum_{k=1}^{m} \gamma_k GAD_{k,\dot{s}(t)} \tag{20}$$

where the number of classes is m, GAD is the Global Adequacy Degree and γ_k is a constant value specified for each class k.

To satisfy $\dot{s}(t) = 0$ with the control action u_{nc} it is necessary to know only the sign of $b(X(t), t)$, to establish the rules based on the classes of LAMDA. In this paper, we have selected $m = 5$, defined as NB: Negative Big, NS: Negative Small, ZE: zero, PS: Positive Small, and PB: Positive Big, values used to establish the rules of the controller. The classes are standardized between $[-1, 1]$, with the centers in $NB = -1$, $NS = -0.5$, $ZE = 0$, $PS = 0.5$, and $PB = 1$.

In (19), if $b(X(t), t) > 0$ we can see that $\dot{s}(t)$ decreases as u_{nc} increases, and vice-versa; thus, the rules that allow obtaining $\dot{s}(t) = 0$ are defined. As an example, if $\dot{s}(t)$ is PB, then a great positive control action u_c is needed to decrease quickly $\dot{s}(t)$. If $\dot{s}(t) = ZE$ (the desired condition), then no control action is required, thus $u_{nc} = ZE$.

Based on the previous analysis, the table of rules corresponding to the control action of the continuous part is proposed in Table 1.

Table 1. Rule table of LSMC for $\dot{s}(t)$ for $b(X(t), t) > 0$

$\dot{s}(t)$				
NB	**NS**	**Z**	**PS**	**PB**
$\gamma_1 = NB$	$\gamma_2 = NS$	$\gamma_3 = ZE$	$\gamma_4 = PS$	$\gamma_5 = PB$

The output u_{nc} must be multiplied with a scaling gain $k_c > 0$ to obtain u_c as: $u_c = k_c u_{nc}$ for calibration.

To compute the discontinuous control action, it is selected the Lyapunov function:

$$V(s(t)) = \frac{1}{2}s(t)^2 \tag{21}$$

The derivative of (21) is:

$$\dot{V}(s(t)) = s(t)\dot{s}(t) \tag{22}$$

For stability, the derivative (22) must meet the condition:

$$s(t)\dot{s}(t) < 0 \tag{23}$$

Replacing (19) in (23), if $u = u_{nd}$:

$$s(t)\dot{s}(t) = s(t)\dot{x}_{dn}(t) - s(t)A(X(t), t) - s(t)b(X(t), t)u_{nd} - s(t)d(t)$$
$$+ s(t)\sum_{i=1}^{n} r_{n-i}\lambda^i e^{(n-i)} < 0 \tag{24}$$

And the control action based on LAMDA considering $O = [\dot{s}(t); s(t)]$ is:

$$u_{nd} = LAMDA_d(s, \dot{s}) = \left| \frac{\operatorname{argmax}(\gamma_k)}{\sum_{k=1}^{m} \gamma_k GAD_{k, \operatorname{argmax}(\dot{s}(t), s(t))}} \right| \sum_{k=1}^{m} \gamma_k GAD_{k, (\dot{s}(t), s(t))} \tag{25}$$

Based on (25), u_{nd} is using for $\dot{s}(t)$ and $s(t)$. In (24), if $s(t)\dot{s}(t)$ is negative for all $s(t) \neq 0$, then the existence of the sliding mode is guaranteed, later the stability analysis is performed.

For the computation of u_{nd}, if $b(X(t), t) > 0$ Table 2 with 25 rules is proposed.

Table 2. Rule table of LSMC for $s(t)$, $\dot{s}(t)$ for $b(X(t), t) > 0$

		$\dot{s}(t)$				
		NB	**NS**	**Z**	**PS**	**PB**
$s(t)$	**PB**	$\gamma_5 = ZE$	$\gamma_{10} = ZE$	$\gamma_{15} = PS$	$\gamma_{20} = PB$	$\gamma_{25} = PB$
	PS	$\gamma_4 = ZE$	$\gamma_9 = ZE$	$\gamma_{14} = PS$	$\gamma_{19} = PB$	$\gamma_{24} = PB$
	ZE	$\gamma_3 = NB$	$\gamma_8 = NS$	$\gamma_{13} = ZE$	$\gamma_{18} = PS$	$\gamma_{23} = PB$
	NS	$\gamma_2 = NB$	$\gamma_7 = NB$	$\gamma_{12} = NS$	$\gamma_{17} = ZE$	$\gamma_{22} = ZE$
	NB	$\gamma_1 = NB$	$\gamma_6 = NB$	$\gamma_{11} = NS$	$\gamma_{16} = ZE$	$\gamma_{21} = ZE$

The output u_{nd} should be multiplied with a scaling gain $k_d > 0$ to obtain u_d as: $u_d = k_d u_{nd}$ for adequate calibration.

In the case that $b(X(t), t) < 0$ only the signs of the γ_k are changed in Tables 1 and 2. Finally, the overall control action of LSMC is:

$$u = k_c LAMDA_c(\dot{s}) + k_d LAMDA_d(s, \dot{s}) \tag{26}$$

The LSMC scheme is shown in Fig. 1. The blocks of the controller for LAMDA are applied in the continuous and discontinuous parts, and the scaling gains in the inputs and the outputs.

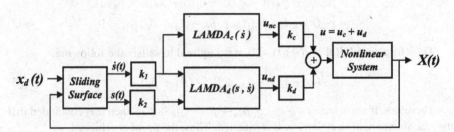

Fig. 1. The control scheme of the LSMC

2.3 Stability Analysis

Rewriting Eq. (24), the Lyapunov function is defined as:

$$\dot{V} = s(t)\left(\dot{x}_{dn}(t) - A(X(t), t) - b(X(t), t)(k_c LSMC(\dot{s}) + k_d LSMC(s, \dot{s})) - d(t) + \sum_{i=1}^{n} r_{n-i}\lambda^i e^{(n-i)} \right) < 0 \tag{27}$$

From (27), it is considered that $\dot{x}_{dn}(t)$ and $\sum_{i=1}^{n} r_{n-i}\lambda^i e^{(n-i)}$ are bounded [14]:

$$|\dot{x}_{dn}(t)| \le \beta_{dn} \tag{28}$$

$$\left|\sum_{i=1}^{n} r_{n-i}\lambda^i e^{(n-i)}\right| \le \beta_e \tag{29}$$

where β_{dn} and β_e are positive constants.

Theorem 1. Considering the system presented in (11), controlled by $u(t)$ in (26), where u_{nc} is in (20), u_{nd} is in (25) and $(k_c + k_d) > \beta_{dn} + \beta_e - (\beta_A + \beta_d)$. Then, the error state trajectory converges to the sliding surface $s(t) = 0$.

Proof. The stability can be analyzed assuming for simplicity $b(X(t), t) = 1$, without loss of generality:

$$\dot{V} = s(t)\left(\dot{x}_{dn}(t) - A(X(t), t) - k_c LSMC(\dot{s}) - k_d LSMC(s, \dot{s}) - d(t) + \sum_{i=1}^{n} r_{n-i}\lambda^i e^{(n-i)}\right) < 0 \tag{30}$$

$$\dot{V} = s(t)\dot{x}_{dn}(t) - s(t)A(X(t), t) - s(t)k_c LSMC(\dot{s}) - s(t)k_d LSMC(s, \dot{s})$$
$$- s(t)d(t) + s(t)\sum_{i=1}^{n} r_{n-i}\lambda^i e^{(n-i)} < 0 \tag{31}$$

From [15] and observing Table 2, it is determined that $k_d u_{nd} = k_d|s|$ and $k_c u_{nc} = k_c|s|$. Thus, replacing (27) in (31), it is obtained:

$$\dot{V} = \beta_{dn}|s| - \beta_A|s| - k_c|s| - k_d|s| - \beta_d|s| + \beta_e|s|$$
$$= \left[-(k_c + k_d) + (\beta_{dn} + \beta_e - \beta_A - \beta_d)\right]|s| < 0 \tag{32}$$

Therefore, to fulfill that $s(t)\dot{s}(t) < 0$, it is required to satisfy the following:

$$(k_c + k_d) > \beta_{dn} + \beta_e - (\beta_A + \beta_d) \tag{33}$$

Therefore, if we select $(k_c + k_d) > \beta_{dn} + \beta_e - (\beta_A + \beta_d)$, then it is concluded that the reaching condition $s(t)\dot{s}(t) < 0$ is satisfied. Thus, the proof is achieved.

2.4 Smith Predictor (SP)

The Smith Predictor is a scheme whose purpose is to compensate systems with dead time. The architecture of the SP is shown in Fig. 2, in which $G_c(s)$ corresponds to the controller, while the transfer function of the process is $G_p(s)$. As can be seen, parallel to the process, there is an approximate model $G_m(s)$ consisting of a stable transfer function $G(s)$ and a dead time t_0. Thus, $G_m(s)$ is defined as:

$$G_m(s) = G(s)e^{-t_0 s} \tag{34}$$

In the structure shown, feedback on the prediction of the process output is performed using a plant model without delay $G(s)$ to improve the system's performance. The difference between the process and model output defined as $e_m(t) = X(t) - X_m(t)$, is fed back to correct modeling errors and disturbances. The closed-loop transfer function is:

$$\frac{X(s)}{R(s)} = \frac{G_c(s)G_p(s)}{1 + G_c(s)G(s) + G_c(s)[G_p(s) - G_m(s)]} \tag{35}$$

If a perfect model of the plant is obtained: $G_m(s) = G_p(s)$, then the closed-loop transfer function reduces to the expression:

$$\frac{Y(s)}{R(s)} = \frac{G_c(s)G_p(s)}{1 + G_c(s)G(s)} \tag{36}$$

This implies that:

$$e_m(t) = X(t) - X_m(t) = 0 \tag{37}$$

Therefore, the feedback consists only of the output of the model without delay. Thus, the dead time is isolated and compensated. However, the SP is not very robust and can present instability due to modeling errors, which would require a robust controller such as LSMC to remain stable over time. Our proposal is abbreviated as LSMC-SP.

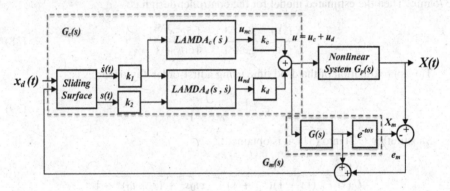

Fig. 2. The control scheme of the LSMC-SP

3 Background Application of the LSMC-SP to a pH Neutralization Reactor

The process used to test the proposed approach is presented in detail in [16]. This system corresponds to a pH neutralization reactor with highly nonlinear characteristics and dominant and variable dead time depending on the operating point. Figure 3 shows the variation of the coefficients over the input signal range [0%–100%], increasing the non-linearity of the system, which is complex to model.

The process can be modeled as a First Order Plus Dead Time (FOPTD) system. For this procedure, the input signal has been varied in the range of 20% to 80% ascending

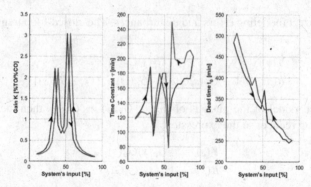

Fig. 3. K, τ, and t_0 varying the input of the system (see [12] for details of the process model) (Color figure online)

and descending, obtaining the variations of the system gain K, the response time τ, and the dead time t_0. Figure 3 shows how these parameters change as a function of the input signal; it varies ascending (blue) and descending (red).

To compute the approximate model of the curves shown in Fig. 3, the variable parameters are averaged, obtaining the following: $K = 1.5$, $\tau = 160$ min and $t_0 = 375$ min. Then the estimated model for the controller design is:

$$\frac{X(s)}{U(s)} = \frac{Ke^{-t_0 s}}{\tau s + 1} = \frac{1.5e^{-375s}}{160s + 1} \tag{38}$$

The approximation of the dead time using a first-order Taylor series is:

$$e^{-tos} \cong \frac{1}{t_0 s + 1} \tag{39}$$

Substituting (32) into (31), it is obtained:

$$\frac{X(s)}{U(s)} \cong \frac{K}{(\tau s + 1)(t_0 s + 1)} = \frac{K}{\tau t_0 s^2 + (\tau + t_0)s + 1} \tag{40}$$

Solving (33) in the time domain:

$$\tau t_0 \ddot{x} + (\tau + t_0)\dot{x} + x - Ku = 0 \tag{41}$$

Equation (34) represented in state-space is:

$$\dot{x}_1 = x_2$$
$$\dot{x}_2 = -\frac{(\tau + t_0)}{\tau t_0}x_2 - \frac{1}{\tau t_0}x_1 + \frac{K}{\tau t_0}u \tag{42}$$

where $x_1 = x$.

The order of the system is $n = 2$; thus, the sliding surface $s(t)$ is defined as:

$$s(t) = \left(\frac{d}{dt} + \lambda\right)^2 \int e(t)dt = \dot{e}(t) + 2\lambda e(t) + \lambda^2 \int e(t)dt \tag{43}$$

From (36):

$$\dot{s}(t) = \ddot{e}(t) + 2\lambda\dot{e}(t) + \lambda^2 e(t) = 0 \tag{44}$$

For $n = 2$ in (14):

$$\ddot{e}(t) = \dot{x}_{d2}(t) - \dot{x}_2(t) \tag{45}$$

Replacing (42) and (45) in (44):

$$\dot{s}(t) = \dot{x}_{d2}(t) + \frac{(\tau + t_0)}{\tau t_0}x_2 + \frac{1}{\tau t_0}x_1 - \frac{K}{\tau t_0}u + 2\lambda\dot{e}(t) + \lambda^2 e(t) = 0 \tag{46}$$

Analyzing Fig. 3, we can conclude that $K > 0$ for the variation of the input signal, then $\frac{K}{\tau t_0} > 0$, so, based on (38), we can conclude that we have the case $b(X(t), t) > 0$. Therefore, Tables 1 and 2 are used for the definition of rules.

Finally, the SP uses the parameters presented in (38) to implement the additional blocks that require the structure shown in Fig. 2.

4 Test and Results

To validate the operation of the proposed controller, this section is presented the comparative analysis of LSMC-SP with other controllers, such as the PID controller, whose parameters have been tuned with the method of Dahlin synthesis detailed in [17], establishing $K_P = 0.297$ and $K_I = 0.033$, and $K_D = 72.79$, the SMC proposed and tuned by the method of [12] has the parameters $\lambda_0 = 0.0000315$, $\lambda_1 = 0.01$, $K_D = 5.35$, $\delta = 0.1$, and for the LSMC, empirically are established the scaling gains $k_1 = 0.01$, $k_2 = 2$, $k_c = 1$ and $k_d = 32$ which are also used for LSMC-SP for a fair comparison.

Simulations validate the tests in two parts: a) disturbance rejection and b) reference changes.

4.1 Disturbance Rejection

Figure 4a shows the response of the controllers when the neutralization process is subjected to a change of -5% in the instant 2000 [min] and of $+5\%$ at the time 8000 [min] and 15000 [min] of $q1(t)$, which is seen as a disturbance in the system.

The results show that SMC presents a highly smooth response, which is why it cannot reach the reference when the system is subjected to disturbances of $\pm5\%$. Also, it can be noted that the PID, LSMC, and LSMC-SP controllers reach the reference similarly in a short time and with a control action (see Fig. 4b) that can be considered adequate because it would not affect the operation of the actuator since it does not have abrupt response intervals.

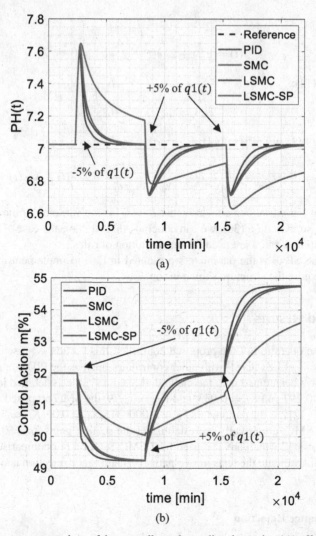

Fig. 4. Performance comparison of the controllers when a disturbance in $q1(t)$ affects the system

Table 3 shows the quantitative analysis of the system responses based on performance indices.

The indices are ISE (Integral Square Error), and TVU (Control Signal Effort) computed for each controller. These results show that the PID is the most aggressive controller in terms of TVU, while the other proposals present similar values of around 120.5.

In terms of ISE, it can be noted that the best proposal is the LSMC-SP, whose ISE value is the minimum (which is the most appropriate) since it is quickly capable of correcting the error and reaching the desired set point.

Table 3. Performance metrics of the controllers for disturbance rejection

Controller	ISE	TVU
PID	1.429×10^4	128.9
SMC	6.992×10^4	120.6
LSMC	2.018×10^4	120.5
LSMC-SP	9.995×10^3	120.5

4.2 Reference Changes

In this simulation, the controllers are tested when the reference changes by -10% at a time of 1000 [min], then from this setpoint, a change of $+33\%$ is introduced to the system at 9000 [min], and finally, a disturbance of $+5\%$ is added at 14000 [min]. Under these conditions, the behavior of the four controllers can be observed in Fig. 5.

Based on Fig. 5a, we can see that the LSMC-SP has the best performance because it quickly reaches the reference, especially when a disturbance is introduced at the last setpoint. The LSMC approach also gets the reference; however, it takes more time and presents a more oscillatory response. As in the previous simulation, the SMC controller again offers a lengthy response and does not reach zero error during the test. Therefore, this proposal requires a recalibration of its parameters.

The PID degrades its performance in the second change of reference because it presents a very aggressive response, making it unstable in the face of the disturbance. In addition, it is not capable of controlling the system. Since the system's operating point has moved, the PID cannot control the process and, therefore, would require a new calibration of parameters, which is not necessary for LSMC-SP, which has proven to be sufficiently robust and with a good disturbance rejection characteristic.

The performance metrics are presented in Table 4. Again, LSMC-SP has the best values because ISE and TVU are the minima.

Finally, the results in Tables 3 and 4 show that the ISE and TVU indices of the LSMC-SP controller are the best, indicating that the control action is smoother (less abrupt) and that the system output reaches the reference more quickly than the other proposals, in addition graphically it is observed that the controller presents very good characteristics of rejection to disturbances.

4.3 Reference Changes Considering Noise Added to the Sensor

In this test, as in the previous section, the plant is subjected to a reference change of $+33\%$ to evaluate the behavior of the controllers. In addition, adding white noise to the sensor signal has been considered, simulating this affectation in the operation of the measuring instrument, with which the controllers' performance and ability to reach the desired reference can be evaluated. The white noise added to the sensor signal is presented in Fig. 6.

The noise shown in Fig. 6 oscillates between values of $\pm 2.5\%$, which generally occurs in real systems; that is why it is opportune to test the behavior of the controllers

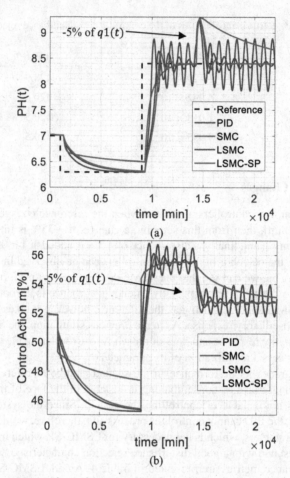

Fig. 5. Performance comparison of the controllers for multiple reference changes and disturbance rejection

Table 4. Performance metrics of the controllers for reference changes and disturbance rejection

Controller	ISE	TVU
PID	2.370×10^5	365.7
SMC	2.759×10^5	156.5
LSMC	2.329×10^5	164.8
LSMC-SP	1.179×10^5	150.2

under these characteristics since, in some cases, the controllers become unstable or generate oscillatory control actions that are not suitable for actuators since they reduce their useful life.

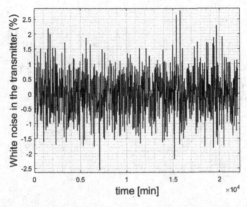

Fig. 6. White noise applied to the sensor signal

Figure 7 shows the system output achieved with each proposal and the respective control action that generates it. Initially, it is observed that the PID is a controller that cannot reach the reference since it maintains an oscillatory behavior around the setpoint but with higher frequency affectations due to added noise which could affect the actuator (the valve in this system) considerably.

The controller SMC has a response that is minimally affected by noise but is very slow in responding to the change of reference and to the disturbance, which, although it is a guarantee for the actuator, not so much for the system since it takes a long time to reach the given setpoint, it should be noted that its control action shown in Fig. 8b is much smoother than the PID.

The LSMC controller presents a quite acceptable behavior since it stays according to what was observed when there is no added noise; however, the noise affects its behavior causing it to act in an oscillatory way to seek to reach the reference (see Fig. 8c).

Finally, the control action of LSMC-PS becomes oscillatory (Fig. 8d) due to the effect of noise in the sensor signal, however, this does not affect it considerably, and the controller does not lose its characteristic of rejection of disturbances and tracking of references, so it can be determined that it is a viable proposal in its practical implementation and that will preserve the useful life of the actuator.

Table 5 shows the rates of the performance indices obtained for the different controllers. In terms of Ise, the LSMC-SP proposal is considerably better than the other proposals, being almost half of the LSMC that is the controller that follows it and much better than the SMC and the PID. In terms of TVU, the SMC proposal is robust against noise; however, as noted in the graphic response, this is because its control signal is very soft and takes a long time to bring the system to the reference. Nevertheless, analyzing this parameter in LSMC-SP is better than the PID and the LSMC, indicating that adding the Smith Predictor and designing the controller for a first-order system is a viable proposal that improves the system's overall performance.

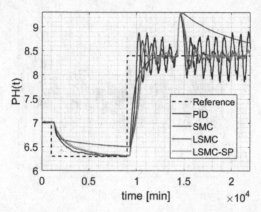

Fig. 7. The output of the system considering noise applied to the sensor signal

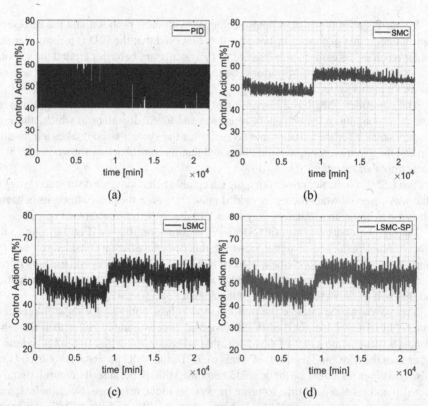

Fig. 8. The output of the controllers considering noise applied to the sensor signal a) PID, b) SMC, c) LSMC, d) LSMC-SP

Table 5. Performance metrics of the controllers for reference changes and disturbance rejection with noise added to the sensor signal

Controller	ISE	TVU
PID	2.370×10^5	4.047×10^4
SMC	2.866×10^5	3.643×10^3
LSMC	2.418×10^5	1.246×10^4
LSMC-SP	1.276×10^5	1.187×10^4

5 Conclusions

In the present work, the LSMC-SP has been proposed to improve the pH process performance. The analysis has shown that adding the Smith Predictor and designing the controller for a first-order system the proposal improves the system's overall performance. The operation of the controller has been validated, and it has been possible to qualitatively observe that the response of the controller is stable for the two simulations carried out, observing a smooth control action that would not affect the actuators and that reaches the desired references quickly even when disturbances are applied. From the quantitative point of view, the results have also shown that LSMC-SP decreases the error more quickly than the other proposals, evidenced by the minimum ISE and a minimum number of oscillations. The proposed controller is effective enough to control the process of variable model parameters and dominant dead time, with excellent disturbance rejection and reference tracking characteristics by a simple design.

Acknowledgment. Oscar Camacho, thanks for supporting this work by the Universidad San Francisco de Quito through the Poli-Grants Program under Grant 1746.

References

1. Bayas, A., Škrjanc, I., Sáez, D.: Design of fuzzy robust control strategies for a distributed solar collector field. Appl. Soft Comput. **71**, 1009–1019 (2018). https://doi.org/10.1016/j.asoc.2017.10.003
2. Denaï, M.A., Palis, F., Zeghbib, A.: Modeling and control of nonlinear systems using soft computing techniques. Appl. Soft Comput. **7**, 728–738 (2007). https://doi.org/10.1016/j.asoc.2005.12.005
3. Knospe, C.: PID control. IEEE Control Syst. **26**, 30–31 (2006). https://doi.org/10.1109/MCS.2006.1580151
4. Morales, L., Aguilar, J., Camacho, O., Rosales, A.: An intelligent sliding mode controller based on LAMDA for a class of SISO uncertain systems. Inf. Sci. **567**, 75–99 (2021). https://doi.org/10.1016/j.ins.2021.03.012
5. Subramaniam, R., Song, D., Joo, Y.H.: T-S fuzzy-based sliding mode controller design for discrete-time nonlinear model and its applications. Inf. Sci. **519**, 183–199 (2020). https://doi.org/10.1016/j.ins.2020.01.010

6. Kaynak, O., Erbatur, K., Ertugnrl, M.: The fusion of computationally intelligent methodologies and sliding-mode control-a survey. IEEE Trans. Ind. Electron. **48**, 4–17 (2001). https://doi.org/10.1109/41.904539

7. Lagrat, I., Ouakka, H., Boumhidi, I., Atlas, B.P.: Fuzzy sliding mode PI controller for nonlinear systems. WSEAS Trans. Signal Process. **2**, 1137–1143 (2006)

8. Morales, L., Aguilar, J., Rosales, A., Pozo-Espin, D.: A fuzzy sliding-mode control based on Z-Numbers and LAMDA. IEEE Access 1 (2021). https://doi.org/10.1109/ACCESS.2021.3105515

9. Estrada, J.S., Camacho, O.: Adaptive sliding mode control for a ph neutralization reactor: an approach based on Takagi-Sugeno fuzzy multimodel, pp. 1–6 (2021). https://doi.org/10.1109/etcm53643.2021.9590774

10. Botía Valderrama, J.F., Botía Valderrama, D.J.L.: On LAMDA clustering method based on typicality degree and intuitionistic fuzzy sets. Expert Syst. Appl. **107**, 196–221 (2018). https://doi.org/10.1016/j.eswa.2018.04.022

11. Sugeno, M., Takagi, T.: Fuzzy identification of systems and its applications to modeling and control. IEEE Trans. Syst. Man. Cybern. **15**, 116–132 (1985)

12. Camacho, O., Smith, C.A.: Sliding mode control: an approach to regulate nonlinear chemical processes. ISA Trans. **39**, 205–218 (2000). https://doi.org/10.1016/s0019-0578(99)00043-9

13. Andaluz, G.M., Morales, L., Leica, P., Andaluz, V.H., Palacios-Navarro, G.: LAMDA controller applied to the trajectory tracking of an aerial manipulator. Appl. Sci. **11**, 5885 (2021). https://doi.org/10.3390/app11135885

14. Roopaei, M., Zolghadri Jahromi, M.: Chattering-free fuzzy sliding mode control in MIMO uncertain systems. Nonlinear Anal. Theory, Methods Appl. **71**, 4430–4437 (2009). https://doi.org/10.1016/j.na.2009.02.132

15. Noroozi, N., Roopaei, M., Jahromi, M.Z.: Adaptive fuzzy sliding mode control scheme for uncertain systems. Commun. Nonlinear Sci. Numer. Simul. **14**, 3978–3992 (2009). https://doi.org/10.1016/j.cnsns.2009.02.015

16. Iglesias, E., García, Y., Sanjuan, M., Camacho, O., Smith, C.: Fuzzy surface-based sliding mode control. ISA Trans. **46**, 73–83 (2007). https://doi.org/10.1016/j.isatra.2006.04.002

17. Smith, C., Corripio, A.: Principles and Practice of Automatic Process Control. Wiley, Hoboken (2006)

Comparison of the Performance of Two Neural Network Models with Parameter Optimization for the Prediction of the Bancolombia Share Price

Juan Sebastian Castillo Amaya, Juan Pablo Ramírez Villamil, and Andrés Eduardo Gaona Barrera[(✉)]

Laboratorio De Automática E Inteligencia Computacional, Universidad Francisco José de Caldas, Bogotá, Colombia
{juscastilloa,jupramirezv}@correo.udistrital.edu.co, aegaonab@udistrital.edu.co

Abstract. The following article explores the characteristics of a NARX neural network, which is used to predict the time series of Bancolombia's stock. The search for the best characteristics of the network is carried out with a heuristic model, which uses a genetic algorithm to vary the different parameters of the network, the solutions explored are 5000 in each independent experiment, out of a total of 15 independent experiments that reduces the probability of falling into local minima.

Keywords: NARX · Forecast · Heuristics

1 Introduction

In 1970, the efficient market hypothesis (EMH) proposed by Fama [1] states that *"it is not possible to systematically predict the prices of a financial asset (bonds, stocks, etc.), since they behave randomly"*. Years later, the hypothesis proposed by Peters [2] of the fractal market (FMH) states that *"prices have a chaotic structure, and could be predicted from nonlinear models, thus rejecting the efficient market hypothesis and invalidating the assumptions of the asset valuation models"* [3].

The works of Suárez [3] and Duarte [4] show that the Colombian market and more exactly Bancolombia's stock corresponds to a fractal market in periods of upswing, therefore, its time series can be predicted by a nonlinear model. In the case of problems in Colombia, neural networks have been used to predict the price of coffee [5], the price of electricity [6] and the reference stock market index of the Colombian Stock Exchange (COLCAP).) [7].

Neural networks have shown better performance in predicting financial assets compared to the traditional Box & Jenkins method, as is the case of [8–11]; The following document addresses the architecture of the neural network NARX used for time series prediction [12–15], mainly focused on the prediction of the

A. D. Orjuela-Cañón et al. (Eds.): ColCACI 2022, CCIS 1746, pp. 99–128, 2023.
https://doi.org/10.1007/978-3-031-29783-0_7

action of Bancolombia. An optimization algorithm is used to tune the NARX and LSTM networks and identify the best features of each, in order to compare their performance in terms of predicting the Bancolombia stock.

2   RNARX Neural Network

A NARX neural network is a time-delayed model that maps an input sequence $(x(t), x(t-1), ..., x(t-m)$ to an output $y(t)$, that is, a feedforward network with input signal delays The NARX architecture has time delay properties but with an autoregressive model, the output to predict is found as a function of the current output and the model inputs [16] Mathematically expressed as:

$$y(n+1) = f(y(n), ..., y(n-q+1), u(n), ..., u(n-q+1))$$ (1)

where f is a nonlinear function of the arguments, $u(n)$ is the current input, $y(n)$ is the current output, and q is time delay units.

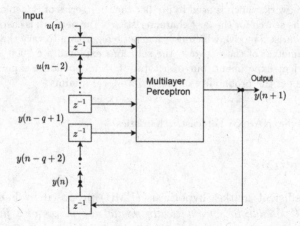

Fig. 1. Representation of neural networks NARX

In the model of the Fig. 1 a delay of q units is applied both in the inputs and in the output, this is applied to the first layer of the multilayer perceptron. The inputs of the model are denoted by $u(n)$ while the output is denoted by $y(n+1)$, which indicates that the output is one unit of time ahead of the inputs. The input values are influenced by the output, so they have exogenous behavior; while the output is a regression of the lagged values themselves.

3   LSTM Neural Network

LSTM networks were introduced by Hochreiter and Schmidhuber (1997) [17] and are designed to counteract the long-term dependency problem more effectively than the GRU model. Unlike the GRU configuration, the memory cells are not the same as the activation functions of the inputs and the use of the gates is done differently.

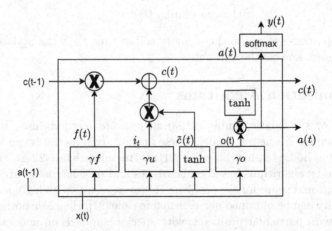

Fig. 2. Module LSTM

As can be seen in Fig. 2, an LSTM module is composed of three types of gates that control the state of the cell: update γu, forget γf and exit *gammaor*. To add or remove information to the module, you must first decide how much information is going to be taken from the previous state through the forget gate. This gate corresponds to a value as a function of the hyperbolic tangent of $c(t-1)$ between 0 and 1 and indicates how much information should be removed from the previous state of the cell. A value of 1 corresponds to keeping the previous state while 0 indicates clearing the state.

In the next step, the update gate is configured to decide which values are updated and a hyperbolic tangent layer for mapping these values. The multiplication of γf and the values $c(t-1)$ is performed, then this result is added to the update gate operation and in this way add the new information. Finally, a last filter is performed to decide what information is relevant to have in the output and then pass it through a function, in this case Fig. 2, softmax and obtain the output $y(t)$.

The memory cell corresponds to the hyperbolic tangent of the matrix model of the input weights and the bias is added.

$$\tilde{c}(t) = \tanh(Wc[a(t-1), x(t)] + b_c) \tag{2}$$

The update gate γu is associated with the candidate to replace in the series, while the forget gate γf is associated with the previous value.

$$\gamma u = \sigma(Wu[a(t-1), x(t)] + bu)$$
$$\gamma f = \sigma(Wf[a(t-1), x(t)] + bf) \tag{3}$$
$$\gamma o = \sigma(Wo[a(t-1), x(t)] + bo)$$

The activation function $a(t)$ is a function of the forgetting gate γo and of the memory cell and the value it takes at time t.

$$c(t) = \gamma u * \tilde{c}(t) + \gamma f * c(t-1)$$
$$a(t) = \gamma o * \tanh(c(t)) \tag{4}$$

where $c(t)$ corresponds to the memory cell at time t, γu the update gate, γf the forget gate, and γo the exit gate.

4   Optimization Algorithms

Automatic or systematic tuning of parameters are methods used to improve the performance of a neural model. These optimization methods can be divided into: exact methods [18], heuristics [19–21] and metaheuristics [22–26]. Unlike the exact methods, algorithms based on heuristics and metaheuristics are not guaranteed an optimal result for the problem. However, results with high performance and reliability can be obtained in a reasonable time [27]. Heuristic optimization is associated with particular problems, with special emphasis on nondeterministic polynomial time (*NP-hard* problems), so its application is very restricted. Metaheuristics, on the other hand, implement high-level algorithms with a higher degree of generalization than heuristics. In this way, these methods are not rooted to specific problems and can be used to optimize different neural networks. The methods that have been developed include: simulated crystallization [25] in which thermal systems and multi-objective optimization are combined, evolution algorithms, among others. One of the most prominent metaheuristics in the literature is the one based on evolution and genetics [24, 26, 28, 29]. This methodology follows a scheme based on variation operators, such as mutation and recombination, and selection operators, such as parent selection and survivor selection, as shown in Fig. 3.

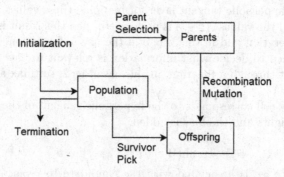

Fig. 3. General scheme of the GA [28]

The parameters of metaheuristic algorithms can be divided into two (2) categories: numerical like the population size in evolution algorithms and categorical like the mutation operator and selection decisions in GA problems [27,28,30]. Likewise, the number of executions for an evolutionary algorithm and the degree of generalization, which can be general or specific [27], are considered.

5   Performance Indicators for Time Series Forecasts

To describe the different performance indicators: $\hat{y}_t$ is the predicted value, y_t is the actual value and n is the size of the test set.

5.1   MAE

The MAE is a difference indicator between two continuous variables. It is an indicator that serves to quantify the distance of error in the points between one series and another.

$$MAE = \frac{1}{n} \sum_{t=1}^{n} |y_t - \hat{y}_t| * 100 \tag{5}$$

5.2   RMSE

The RMSE measures the root mean square deviation of the predicted values. This indicator is sensitive to changes in scale and transformations of the data, penalizes extreme errors and does not provide information about the general direction of the error [31].

$$RMSE = \sqrt{\frac{1}{n} \sum_{t=1}^{n} (y_t - \hat{y}_t)^2} \tag{6}$$

5.3   POCID

The POCID represents the number of hits in the prediction when the value of the time series decreases or increases.

$$POCID = \frac{\sum_{i=1}^{n} D_i}{n} \tag{7}$$

where:

$$D_i = \begin{cases} 1 \; si \; (y_t - y_{t-1})(\hat{y}_t - \hat{y_{t-1}}) > 0 \\ \\ 0 \; si \; (y_t - y_{t-1})(\hat{y}_t - \hat{y_{t-1}}) \leq 0 \end{cases} \tag{8}$$

The value of POCID takes the range $\{x/0 \leq x \leq 1, x \in \mathbb{R}\}$, and the closer to 1 the better the prediction model. This indicator is important as it measures the change in direction, which in the case of stock volatility represents trend changes that affect profit and loss.

5.4 Coefficient of Determination R^2

The coefficient of determination is the proportion of the sample variance of the variable explained by the prediction. The value of this indicator is between 0 and 1 and is dimensionless. The coefficient of determination is defined as:

$$R^2 = 1 - \frac{\sum(\hat{y} - y)^2}{\sum(y - \bar{y})^2} \tag{9}$$

where $\hat{y}$ corresponds to the value of the prediction and $\bar{y}$ the mean of the y values. This indicator reflects the goodness of the model with respect to the variable to be predicted, the closer its value is to 1, the greater the fit to the variable to be predicted and therefore greater reliability.

6 Proposal for Neural Network Tuning

The proposed model consists of six (6) stages: the time series, the input characteristics, the neural model NARX or LSTM, the genetic algorithm, the evaluation of the best models and the comparison of the best models. Best results. The block diagram of the proposed model is presented in Fig. 4:

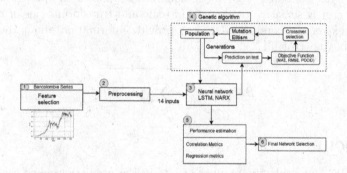

Fig. 4. Proposed model for the prediction problem

6.1 Bancolombia Series

The Bancolombia share database that is reported in the BVC from January 1, 2001 to December 31, 2018 is used for training, validation and testing of the model proposed in Fig. 4. This time interval is equivalent to a total of 4336 d of operation, whose closing prices are represented in Fig. 5.

Table 1. Technical indicators used as inputs in the model

Technical Indicators

Name	Description	Period (days)
ADX	It uses the maximum, minimum prices and an accumulated period.	14
Momentum	Difference between the current price of the series and the previous price.	1
MACD	It uses two sma on the closing price and one EMA for the MACD signal.	SMA: 12 SMA: 26 EMA: 9
RSI	It uses the calculation of the average of the gains, losses.	14
Stochastic K	Maximum, minimum and closing price.	14
Stochastic D	The values of the stochastic K in an accumulated period.	7
ADO	It uses prices and closing volume.	1
DPO	The closing price and the moving averageof it.	20
DEMA	Two exponential moving averages one being the mean of the other.	30
Bollinger Bands	Upper and lower band with respect to the moving average.	20
OBV	Volume flow to predict price changes.	1
NVI	Price ranges and volume.	1
CCI	Comparison between current price with respect to the average.	1

Fig. 5. Bancolombia share closing price from 2001-01-04 to 2018-12-28

1) Technical analysis: Technical analysis consists of extracting behaviors and characteristics of the time series from the calculation of technical indicators.

According to the indicators presented in [32] and with the information provided in the BVC of price range and volume, the 14 characteristics used as inputs to the model are chosen. The description and time periods used for these indicators are presented in Table 1.

6.2   Data Preprocessing

Once the technical analysis has been carried out on the series, the set obtained corresponds to the 14 indicators together with the closing price in the 4336 sessions. Since the different indicators have periods for their calculation, it is necessary to modify the data set to avoid NAN data; the data set number results in 4278 rows and 15 columns. The data is divided into training, validation and test sets as follows:

- Training set: 3421 records which corresponds to 80% of the total set from April 30, 2001 to December 31, 2015.
- Validation set: 715 records corresponding to 16.7% from January 1, 2016 to June 31, 2018.
- Test set: 141 records which corresponds to 3.3% from July 1 to December 28, 2018.

Once the data is organized, it is normalized to accelerate the learning process and reduce the wide range of values that the closing price of the share can take [33]; to perform the normalization, the minimum and maximum method of the *sklearn* tool is used. This process is done for all three (3) data sets; taking into account the same normalization parameters in training and test to avoid bias in the latter set. The mathematical expression of this normalization is shown as:

$$x_{normalized} = \frac{x - min}{max - min} \tag{10}$$

where the normalization is performed for each column of the database so that $x_{normalized}$ corresponds to the new normalized value, x the value to normalize, min is the minimum value of the column while max the max.

6.3   NARX Network

For the construction of the NARX network, the resources of the *PyTorch* library are used for the construction of the neural network and the *Sysidentpy* library to perform the feedback with the delay, the diagram of the network can be seen in Fig. 1.

6.4   LSTM Network

For the construction of this LSTM network, the *Tensorflow* library is used together with the *Keras* dependency, tools used for this type of model and with a lot of documentation. It is worth mentioning that in order to obtain a better result, bidirectional LSTM networks are used, this type of network performs the training in direct sequence of the series and then in reverse sequence.

These models, being recurring, admit as input vectors of the form (*batchSize*, *TimeSteps*, *features*), which correspond to the size of the experimental batch, the time step in each sample and the number of features that are equal to the fourteen (14) entries.

6.5   Genetic Algorithm

In order to establish a broad search for architectures and suitable parameters for the NARX and LSTM models, a genetic algorithm based on the evolution of individuals, selection, crossing, mutation and elitism is performed. Figure 6 shows the process of this algorithm, where there are five (5) parameters that must be defined a priori, namely the maximum number of search generations, the population size of each generation, the number of parents per generation, the number of elites per generation and the rate of mutation of individuals, these five parameters must be defined by the user.

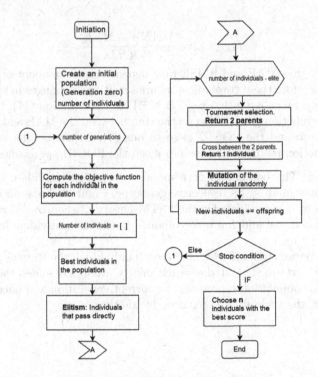

Fig. 6. Genetic Algorithm Flowchart

1) Genotype. The genetic algorithm establishes the genotype of each individual as the first fundamental aspect, which allows for mutation, reproduction, selection and the descent of the individuals in the population. A genotype is a collection of genes grouped in a chromosome where each gene corresponds to the parameters of interest to vary in each neuronal model. The genotype of an individual NARX can be seen in Fig. 7 and in Fig. 8 the genotype of the LSTM network.

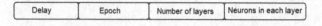

Fig. 7. Genotype of an individual NARX

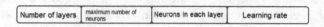

Fig. 8. Genotype of an individual LSTM

2) Objective Function. For the evaluation of each model, an objective function is established that depends on three (3) performance indicators: MAE, POCID and RMSE.

$$F_{Objetivo} = \frac{MAE + RMSE}{POCID} \tag{11}$$

The range that MAE and RMSE take depends on the amount of data to be evaluated, therefore these two indicators are sensitive to changes in scale, unlike POCID; whose range is $\{x/0 \leq x \leq 1, x \in \mathbb{R}\}$. In the equation (11), the value of the objective function gets smaller as the sum between the MAE and the RMSE approaches zero and the POCID close to one (1). Otherwise the value of the objective function tends to large values when the POCID approaches zero (0).

3) Population. The population corresponds to a set of candidate individuals, each one represented by its respective genotype. Figure 9 shows an example of a population of n individuals for the LSTM network, where *Ind 1* corresponds to the first individual and *Ind n* corresponds to the last individual in the entire population.

Once the parameters are assigned to each individual, the training, validation and test are performed, and then each one is evaluated under the objective function. Each population represents the current generation and once it evolves through time, the generation is replaced by another.

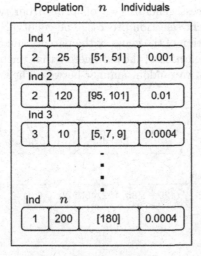

Fig. 9. Population of n individuals represented by chromosomes

4) Selection. Once the population is evaluated by the objective function, it must be decided which individuals participate in the cross to produce the offspring. There are various selection methods such as: roulette method, ranking and tournament competition; all based on the order of the individuals according to their performance. In this case, selection by tournament is used since it generates less discrimination than the other methods.

In Fig. 10 shows an example of tournament selection. First, two pairs are selected from the population where individuals one (1) and two (2) correspond to the first pair and individuals three (3) and four (4) correspond to the second. From each couple, the best individual is selected, represented by the winner one (1) and two (2) respectively. The winners compete with each other according to their objective function to generate the father of the offspring.

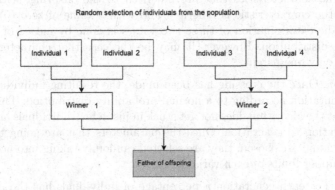

Fig. 10. Competitive tournament selection method

5) Crossing. This process generates an offspring from two parent individuals chosen in the tournament selection process. The genotype of the new individual is the random combination of the genes of the parents and can be done in various ways. In this case, the selection of the parameters that the new individual inherits is carried out in a uniform random manner between the two parents.

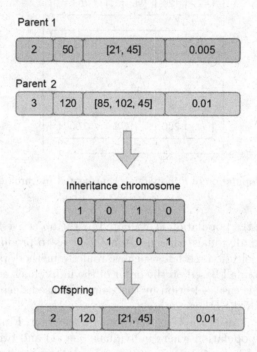

Fig. 11. Crossing of individuals to generate offspring

In Figure 11 the crossover is uniform between the two parents, parent one (1) represented in red and parent two (2) in blue. The inheritance chromosomes represent the characteristics that are inherited to the offspring, with a value of one (1) the characteristic is inherited and with a value of zero (0) it is not inherited; the determination of these two values is done by means of a normal probability distribution. However, it may be the case that characteristics are defined by only one parent.

6) Mutation. Once the crossing has been made, the resulting individual is subjected to mutation according to a normal probability distribution. This process ensures that the algorithm does not get stuck in just a few individuals and change from generation to generation. Once the parameters that are going to mutate in the individual are chosen, they are selected randomly taking into account the lower and upper limits of each variable.

7) Elitism. For each generation a percentage of individuals has the possibility of passing directly to the next generation. These individuals are the ones that present the best results in the objective function and are called elites, through this process it is ensured that the population does not fall into local minima and that each generation is better than the previous one.

Figure 12 shows an example of a population of ten (10) individuals with an elitism of 0.1, which indicates that an individual (10% of the population) from generation one (1) passes directly to generation two (2). The remaining individuals are the product of selection, crossbreeding, and mutation.

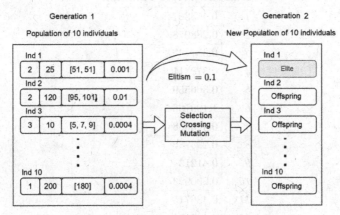

Fig. 12. Process to create a new generation

7   Experiments and Results

7.1   NARX Network

Experimentation with the NARX network is performed only with the NARX optimizer since no significant changes were observed in the objective function when using *Adadelta, Adagrad, ASGD* or *RMSprop*.

The characteristics that the experimentation has are defined in the Table 2, where the parameters of GA are to define how the search for characteristics will be carried out and the parameters NARX are to delimit the search surface.

Table 2. Characteristics of experimentation by objective function of independe Runs

Characteristics of independent runs			
GA parameters		**NARX parameters**	
Maximum of generations	100	*Neurons per layer*	1 − 90
Population size	50	*Layers for the model*	2 − 250
Number of parents	5	*Epoch for training*	100 − 300
Elites by generation	1		
Mutation rate	10%		

Table 3 shows in order the 15 runs made in the NARX experiments and the value of the objective function of the best individual of each run. Each of these tests has 5000 individuals, for a total of 75000 neural networks evaluated in the experimentation.

Table 3. Narx Network Independent Runs

Run	Objective Function
1	0.403367
2	0.288078
3	0.357148
4	0.378311
5	0.366559
6	1.199196
7	0.400688
8	0.295786
9	0.41213
10	0.290923
11	0.486715
12	0.441171
13	0.448631
14	0.412526
15	0.428646

Figure 13 shows the distribution of the error in the objective function of the best individuals, in each generation (100) for the 15 independent runs, which gives us a total of 1500 data; In the graph it can be seen that 50% of the individuals are found in the first three bars, which implies that half of the networks found by GA have an error equal to or less than 0.54 in terms of to objective function.

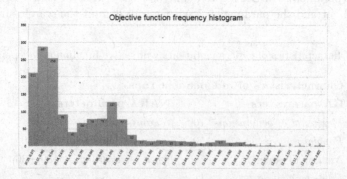

Fig. 13. Elites of each generation of the 15 independent races

In order to demonstrate the independence of the independent runs, the chi-square test is performed for the first three runs, which starts from the null hypothesis that the variables are independent. Since the p-value in the Table 4 is greater than 0.05, it is not possible to reject the null hypothesis and therefore there is no dependency association between the runs.

Table 4. Chi square test

Chi square test		
Statistical	*p-value*	*df*
4.9234	1.00	198

For the Table 5 the five (5) best individuals are selected by objective function of the independent runs, where it can be seen that the output delays tend to converge to two (2) and the GA tends to find to the best individual before half of the total generations, but the sample of five (5) individuals is too small to assert that this is a general rule. The individual with the best objective function is the one with the fewest number of neurons and layers, which suggests that larger networks are not necessarily better for predicting the Bancolombia series.

Table 5. Best individuals by objective function of independent runs

	Best Individual Independent runs				
	1	2	3	4	5
Run	2	10	8	3	5
Generation	40	36	13	17	2
Layers	14	240	76	235	186
Neurons	648	10759	3732	10590	8535
Delays	2	2	2	3	2
Epoch	128	131	130	221	278
Objective Function	0.288078	0.290923	0.295786	0.357148	0.366559

Figure 14 presents the range of neurons explored in the five (5) best independent runs, in the form of a box-whisker plot. The boxes are arranged according to the performance of each run from left to right. Each box shows the range of neurons used by the 5000 individuals of each test, in boxes four (4) and five (5) it can be seen that most of the individuals are above 4000 neurons and that their exploration is less in terms of range of neurons with respect to runs one (1), two (2) and three (3).

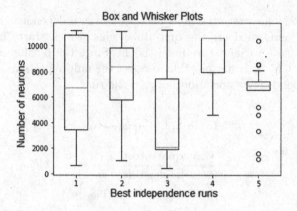

Fig. 14. Whisker plot of the number of neurons per run in the top five runs

In the Table 6 it can be seen that the three best individuals by objective function have values very close to each other, with differences of 0.001 in the MAE and in the RMSE; for the POCID the values are not more than one percentage point away from the 50% prediction of the direction of the time series.

Table 6. Performance indicators of the best individuals in the experiment

	Best individuals of experimentation		
	Regression metrics		
Run	MAE	POCID	RMSE
2	0.029600	0.5	0.114438
10	0.031016	0.507142	0.116523
8	0.030274	0.492857	0.115505
3	0.037812	0.5	0.140762
5	0.048851	0.492857	0.131809

The prediction of the best individual is plotted together with the real data in Fig. 15, where it can be seen that the initial data is very far from the real value. Due to this, a prediction approach is made without taking into account the three (3) initial values as shown in Fig. 16. These three (3) initial data points increase the error of the objective function by 0.21, which is 72% of the total error obtained with the 141 data points of the prediction window. The difference in the value of the objective function is given mainly by the magnitude of the MAE and RMSE errors, since the POCID remains at 0.5 for both cases. The difference regarding the MAE is 0.014173 while for the RMSE it is 0.0946.

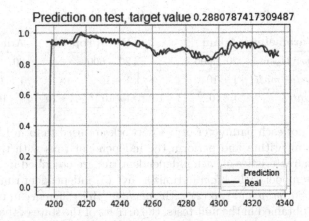

Fig. 15. Best individual of the Independent runs Test

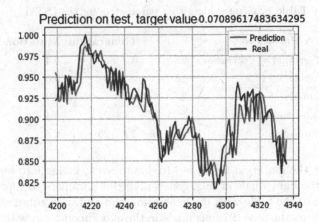

Fig. 16. Close Up Best individual Of The Independent Runs Test

7.2 LSTM Network

Experimentation for LSTM networks is divided into three types of tests, according to the optimizers: *Adam*, *RMSprop*, and *Adadelta*; which showed remarkable results in works like [34] with *Adam* and [35] with *Rmsprop*.

In the first instance, unit tests are carried out with each type of model in order to establish a range of values for the network parameters and with values accepted in [34–36]. The result of these allows narrowing the search space and obtaining more tests with accepted ranges, especially in LSTM networks that present greater complexity in their architecture and longer training time than *feedforward* networks. Table 7 shows the range of parameters for each optimizer.

Table 7. Parameters of the LSTM models

LSTM Parameters	Adam	RMSprop	Adadelta
Neurons per layer	20 – 600	20 – 600	20 – 800
Layers for the model	1 – 10	1 – 10	1 – 15
Learning rate	$1 * 10^{-5} - 1 * 10^{-2}$	$1 * 10^{-4} - 1 * 10^{-2}$	$1 * 10^{-3} - 1$

The ranges of each parameter of the network are used to form the genotype of the genetic algorithm and perform the independent runs with the proposed model. The characteristics of each independent run are described in Table 7. For the case of the model with *Adam* optimizer, five (5) independent runs are made, two (2) with *RMSprop* and two (2) with *Adadelta*; this difference in runs is given by the results obtained in the unit tests, the articles of the state of the art [34–36] and the limitations in terms of time and hardware for the LSTM models.

Table 8. Genetic algorithm parameters for the LSTM

GA Parameters		Generation	Mutation
Maximum of generations	25	1 – 8	0.4
Population size	15	8 – 13	0.6
Number of parents	2	14	0.9
Elites by generation	1	15 – 22	0.4
Mutation	Variable	22 – 25	0.8

In Table 8 the mutation factor is variable to avoid falling into local minima and to vary the individuals in each generation; this parameter is represented on the right side of the table with the generation and mutation columns.

Table 9 shows the nine (9) runs made in the experimentation with the LSTM and the value of the objective function of the best individual of each run. The first five (5) of these runs are done with the *Adam* optimizer, two (2) with *RMSprop*, and two (2) with *Adadelta*.

Table 9. Independent runs of the LSTM network

Optimizer	Run	Objective Function
Adam	1	0.068484
	2	0.071566
	3	0.065777
	4	0.075145
	5	0.066653
Rmsprop	6	0.083952
	7	0.067291
Adadelta	8	0.076324
	9	0.071942

To demonstrate the independence of the runs, a chi square test is performed between 2 runs of the algorithm. This test starts from the null hypothesis in which the variables are independent. According to the Table 10 the significance value is greater than 0.05 and therefore the null hypothesis cannot be rejected, in this way there is no dependency association between the runs.

Table 10. Chi square test

Chi square test		
Statistical	*p-value*	*df*
19.84	0.9954	39

Figure 17 shows the frequency distribution of the objective function of the 5 best individuals of each generation for 2 independent runs. The graph shows the largest distribution of the objective function between the values of 0.07 and 0.08, with an approximate representation of 27%. In addition, they present outliers concerning architectures with poor performance beyond the value of the objective function of 0.25.

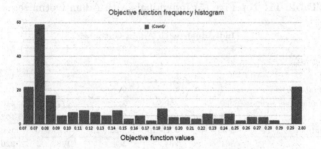

Fig. 17. Top 5 individuals from each generation of 2 separate runs

Figure 18 represents the objective function throughout the 25 generations in run 3. As can be seen in the first 10 generations, there is a high variance between the data, resulting in greater fluctuations. When passing to generation 14, the error in the objective function increases due to the mutation rate of 0.9 and that is necessary to give greater exploration, to then stabilize in the last 5 generations and present less variance.

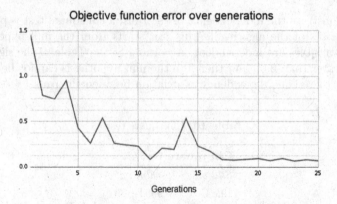

Fig. 18. Behavior of the error during the third run

1) LSTM network with Adam optimizer. Of the five (5) runs made with the *Adam* optimizer, the best individual of each run is extracted. The table 11 presents these individuals and their parameters arranged from left to right according to the value of the objective function.

Table 11. Top Five (5) Individuals with Adam Optimizer

	Individuals with Adam Optimizer				
	1	**2**	**3**	**4**	**5**
Independent Run	3	5	1	2	4
Generation	2	5	3	8	15
Number of layers	2	1	2	2	1
Maximum number of neurons	277	250	312	277	416
Neurons	$190 - 195$	239	$261 - 242$	$190 - 195$	137
Learning rate	$3.25 * 10^{-5}$	$1.1 * 10^{-5}$	$1.25 * 10^{-5}$	$3.25 * 10^{-5}$	$1.49 * 10^{-5}$
Epoch	364	364	364	168	320
Objective Function	0.065777	0.066653	0.068484	0.071565	0.075145

The best results for the architecture of the neural network is one (1) or two (2) layers and with a number of neurons between 130 to 250 as shown in Table 11. The learning rate is in the order of 10^{-5}, so a small value is preferably observed in this parameter. Figure 19 represents the prediction on the test data of the best individual of all runs with the optimizer *Adam*, found in the second generation of the third run. This figure shows the value of the objective function and the real time series.

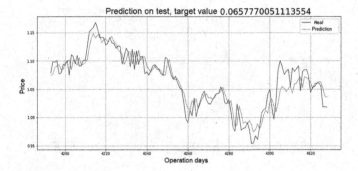

Fig. 19. Performance on unknown data Individual 1 whit ADAM Optimizer

As can be seen in Fig. 19, the prediction follows the trend of the series with an objective function value of 0.06577. The MAE between the actual and predicted signal is 0.01423, the RMSE is 0.01840, and the POCID is 0.4962. Of the three (3) performance indicators mentioned, the MAE presents the lowest value with a difference of 0.004 over the value of the RMSE, this indicates that there is a smaller error in the distance between the points of the real series and of prediction. On the other hand, the POCID of 0.49 indicates errors in the exact direction points, in percentage terms approximately 50% of the prediction points follow the direction of the real series; this is evidenced by the prediction delay at the beginning and end of the prediction.

2) LSTM network with RMSprop optimizer. From the two (2) runs made with the *RMSprop* optimizer, the five (5) best individuals are extracted. Table 12 presents these individuals and their parameters arranged from left to right according to the value of the objective function.

Table 12. Top five (5) individuals with Rmsprop optimizer

	Individuals with RMSprop Optimizer				
	1	2	3	4	5
Independent Run	2	2	1	1	1
Generation	6	3	9	14	18
Number of layers	1	1	3	1	2
Maximum number of neurons	249	104	249	249	128
Neurons	90	88	123 − 167 − 247	18	94 − 85
Learning rate	$3.81*10^{-5}$	$5.32*10^{-5}$	$3.36*10^{-5}$	$2.53*10^{-4}$	$5.32*10^{-5}$
Epoch	364	364	364	364	364
Objective Function	0.067291	0.0686220	0.083952	0.085780	0.098863

The best results for the architecture of the neural network are one (1), two (2) and three (3) layers, with neurons between 18 to 240 as shown in Table 12. The learning rate is small on the order of 10^{-5} except for individual four (4). Figure 20 represents the prediction on the test data of the best individual of the runs with the optimizer RMSprop, found in the sixth generation of the second run. This figure shows the value of the objective function and the real time series.

Fig. 20. Performance on unknown data Individual 1 whit RMSprop Optimizer

For the prediction of Fig. 20 the value of the objective function is 0.065777 with a difference of only 0.00152 with that of the best optimizing individual *Adam*. The MAE for this prediction is 0.015021, the RMSE is 0.01987 and the POCID is equal to 0.518518, with these indicators only the pocid exceeds by 0.022318 the result obtained with the optimizer *Adam*.

3) LSTM network with Adadelta optimizer. From the two (2) runs carried out with the Adadelta optimizer, the five (5) best individuals are extracted. Table 13 presents these individuals and their parameters organized from left to right according to the value of the objective function.

Table 13. Top five (5) individuals with Adadelta optimizer

	Individuals with Adadelta optimizer				
	1	**2**	**3**	**4**	**5**
Independent Run	2	1	1	2	2
Generation	6	5	7	13	8
Number of layers	1	2	1	1	1
Maximum number of neurons	392	431	275	392	354
Neurons	357	288 – 431	206	357	306
Learning rate	0.365934	0.349164	0.349164	0.085429	0.035061
Epoch	364	364	364	364	364
Objective Function	0.071942	0.076324	0.086134	0.086701	0.108004

The best results for the architecture of the neural network are 1 and 2 layers, with neurons between 200 and 357 as shown in Table 13. The learning rate is higher than that observed in the other optimizers with a range between 0.035 to 0.36. Figure 21 represents the prediction on the test data of the best individual of the runs with the optimizer *Adadelta*, found in the sixth generation of the second run. This figure shows the value of the objective function and the real time series.

Fig. 21. Performance on unknown data Individual 1 whit Adadelta Optimizer

The value of the objective function for the prediction of Fig. 21 is 0.071942 with a difference of 0.004652 and 0.006178 with the best individuals of *RMSprop* and *Adam* respectively, therefore it presents the lowest performance in terms of objective function. The MAE for this prediction is 0.0153256, the rmse is 0.0203 and the pocid is equal to 0.496296, the latter is equal to the result of the optimizer *Adam* while the MAE and RMSE present the highest values.

4) Selection of the final individual LSTM. According to the results obtained for the three types of optimizers *Adam*, *RMSprop* and Adadelta, the final LSTM individual with the best performance is the first in the Table 11 with a value objective function of 0.065777, found in the second generation of the third run with the optimizer *Adam*. Figure 22 represents the architecture of this individual.

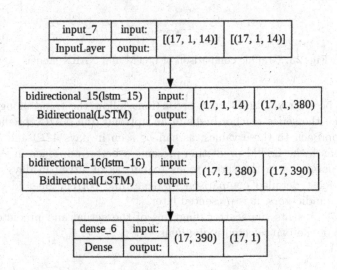

Fig. 22. Architecture of the best individual LSTM

In Fig. 22 the input layer receives a vector of the form (*batchSize*, *timeSteps*, *features*) with the value of (17, 1, 14). The second layer is bidirectional with 190 neurons and 380 outputs due to the processing of the series in direct and inverse sequence. The third layer receives the outputs of the previous layer with the parameter *ReturnSequences = True*, to return 390 outputs. The output layer is a dense layer of a neuron whose output is a vector of the form (17,1), as in Fig. 22 is dense since it gathers all the outputs of the layer previous.

7.3   Comparison Between LSTM and NARX Networks

Once the best individual for the two networks has been identified, these are compared based on the performance of the prediction with test data and the complexity of their architecture. For a comparison consistent with the initial time series, the forecast data is denormalized and compared with the actual test series; as presented in Fig. 23. In this figure, an approach to the time series is made to avoid visualizing the large errors of the NARX network at the beginning of the prediction and in this way to observe the most important points of the figure.

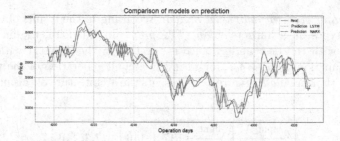

Fig. 23. Graphic comparison of LSTM and NARX models

As can be seen in Fig. 23 both predictions follow the trend along the real series. When the series reaches highs in an uptrend, the NARX prediction has a better approach to these values; as can be seen in days 4210 and 4310. On the other hand, the LSTM prediction presents a better approach in those local maximum points when the series tends to fall, as observed in days 4250 and 4270. The most detailed comparison in Regarding correlation and regression performance indicators, it is presented later.

Figure 24 presents the scatter diagrams of the actual and predicted values for the test of the two selected individuals.

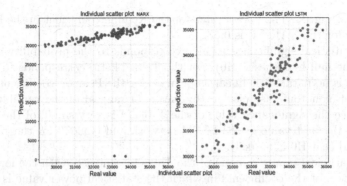

Fig. 24. Dispersion of LSTM and NARX individuals

In Fig. 24 it is observed that·the dispersion of the NARX individual presents three (3) atypical values in the group of points of the diagram, this error is also shown in Fig. 15. For the case of the LSTM network scatter diagram, the points are grouped uniformly with a positive slope; therefore, the linear correlation is greater than that shown by the NARX network, as indicated in Table 14.

Table 14. Correlations of the best individuals of the experimentation where r corresponds to the correlation coefficient, CI95% the confidence interval of 95%, p-val significance and power the statistical power.

		Correlation measures			
		r	CI95%	p-val	power
LSTM	*Pearson*	0.927412	[0.9, 0.95]	4.37554*10–59	1.0
NARX		0.2270	[0.06, 0.38]	0.00677	0.7734
LSTM	*Spearman*	0.917271	[0.89, 0.94]	1.98325*10–55	1.0
NARX		0.8198	[0.76, 0.87]	1.78*10–35	1.0
LSTM	*Kendall*	0.762531	[0.68, 0.82]	1.93698*10–39	1.0
NARX		0.66308	[0.56, 0.75]	2.56*10–31	1.0

In Table 14 the Spearman and Kendall correlation coefficients improve in the NARX network by 0.5928 and 0.4360 respectively compared to the linear coefficient, this is due to the fact that these coefficients are less sensitive to the outliers that are observed. On the scatterplot. For both individuals, the Kendall coefficient reports lower values than the Spearman coefficient. In the case of LSTM, this difference is 0.1546, while for NARX it is 0.1567.

The CI95% confidence interval of the coefficients indicates the range of values between which the correlation coefficient is found with a 95% reliability. For the case of the LSTM in the Table 14 the minimum value in its linear coefficient is 0.9 while for the NARX it is 0.06; difference that is noticeable in the scatter diagrams of Fig. 24. The lowest value for the Kendall coefficient in the LSTM is 0.68 while for the NARX it is 0.56.

The statistical significance p-value corresponds to the probability of fulfillment of the null hypothesis (H0). For the Table 14, H0 corresponds to the fact that there is no parametric linear association for the Pearson coefficient or nonparametric monotonic association (by means of ranges) of the variables under study. Since the significance value of the Table 14 is very small, of the order of 10^{-59} for the best value of the LSTM network and 0.00677 for the NARX, it can be said that H0 is unlikely.

The statistical power (power) of the Table 14 contrasts with the low significance value. For the coefficients in which the statistical power value is one (1), H0 can be rejected under the premise that this hypothesis is false. In the case of NARX, the value of 0.7734 in this indicator does not give total reliability to reject H0 and therefore the linear relationship between the prediction and the real values can be questioned.

Table 15 presents the regression metrics in terms of coefficient of determination (R2), MAE, POCID and RMSE for the best LSTM and NARX individuals.

Table 15. Performance indicators of the best individuals in the experiment

| | Best individuals of experimentation | | | |
| | Regression metrics | | | |
	R2	MAE	POCID	RMSE
LSTM	0.858603	0.0142387	0.4962962	0.018406
NARX	-6.8604	0.0291542	0.5012151	0.114438

In Table 15, the NARX determination coefficient with a negative value of -6.8604 indicates that the prediction made does not describe the test data series in terms of sample variance. The Eq. 9 of the coefficient of determination of the Sect. 5.4 shows that the values of the coefficient can become so negative the greater the difference between the predicted and real values, these points of greater variance correspond to the atypical values that are observed in the dispersion diagram of Fig. 24. In the case of R2 for the LSTM, the value of 0.8596 is close to 1 and therefore represents a greater adjustment to the real values.

As presented in Table 15, the MAE of the LSTM is lower than that presented with the NARX network with a difference of 0.01491. The POCID on the other hand remains the same for both models with a performance of approximately 0.5. The RMSE remains very small for both individuals in the order of 0.114438 and 0.018406.

The Table 16 presents the parameters of each individual according to its architecture and the training configuration. For the architecture, the number of layers, number of neurons and delays are presented; Regarding the training configuration, we have: the optimizer used, the learning rate and the epochs.

Table 16. Parameters of the best individuals of the experimentation

| | Network Comparison | | | | | |
Network	# Layers	# Neurons	Optimizer	Learning rate	Epochs	Delays
LSTM	2	385	Adam	$3.25 * 10^{-5}$	364	1
NARX	14	648	Rmsprop	1.0	128	2

Como se presenta en la Table 16 la red NARX utiliza 263 neuronas y un retardo más que la red LSTM. A pesar de esta diferencia los módulos que componen la red LSTM presentan una mayor complejidad que los nodos *feedforward* de la red NARX. Por otro lado en la configuración del entrenamiento la red LSTM tiene mas épocas que la red NARX con un tasa de aprendizaje muy pequeña; esto influye de igual manera en el tiempo de entrenamiento de la red donde un individuo LSTM toma aproximadamente 17.7 minutos y un individuo NARX 50 segundos.

As shown in Table 16, the NARX network uses 263 neurons and one more delay than the LSTM network. Despite this difference, the modules that make up the LSTM network are more complex than the *feedforward* nodes of the NARX network. On the other hand, in the training configuration, the LSTM network has more epochs than the NARX network with a very small learning rate; this equally influences the network training time where an LSTM individual takes approximately 17.7 min and a NARX individual 50 s.

8   Conclusions

The present work was developed in search of a tool to improve the price predictions in the BVC of a share, it was decided to use the Bancolombia share since it is one of the most relevant banks in the country and contributes an important part to the economy of it. During the investigation of the state of the art of the computational tools used in financial series prediction problems, the LSTM and NARX neural networks were found. Due to the fact that the performance of the networks in the prediction of this time series is not known a priori, it was necessary to develop a model, in which the following is carried out: extraction of characteristics of the action, data pre-processing, construction of the networks, genetic algorithm for tuning of the parameters and the comparison under performance indicators.

The extraction of characteristics carried out in the model consists of the calculation of technical indicators that describe the behavior of the series, then the max-min normalization of the data and division of sets in training, validation and test of the networks is carried out. The genetic algorithm used for the tuning of the networks has the particularity of implementing a selection by double tournament, thereby reducing the discrimination of characteristics and the probability of falling local minima at the time of performing the exploration. This algorithm also has adaptive mutation to perform a broad search with a population of 15 individuals, which can be considered small, as is the case with the

LSTM network. The objective function that is used to evaluate each individual of the genetic algorithm consists of the MAE, RMSE and POCID indicators with the purpose of evaluating the magnitude of the error and the direction of the series.

Experimentation with neural networks was carried out in parallel on different computing machines. In the case of the LSTM network, nine (9) independent runs were carried out with 25 generations and 15 individuals, each run lasting an average of 75 h with a total of 28 d of experimentation. In the case of the NARX, 15 independent runs were carried out with 100 generations and 50 individuals, each run lasting an average of 41.6 h with a total of 26 d of experimentation. However these times depend on factors such as the architecture of the individual, the hardware and software used and additional times to save the resulting data. The difference in total individuals evaluated, 3,375 for the LSTM and 75,000 for the NARX, is clearly due to the difference in training times and the complexity of each model; On the one hand, the LSTM network is bidirectional and has modules that have a variety of control parameters, while the NARX manages nodes *feedforward* which are unidirectional. In addition to the complexity of the network, the training of the LSTM is carried out in an average of 364 epochs, unlike the 128 of the NARX.

In the case of the LSTM network, the best individual in terms of objective function performance was the ADAM with a value of 0.065777. In the case of the NARX network, the best individual had a target value of 0.28807, this high error was due to the three (3) outliers that were present at the beginning of the prediction. Without these points, the value of the objective function improves to 0.0708 with a difference of 0.005032 compared to the LSTM individual. In the coefficient of determination there is a difference of 0.85 between both individuals so that the LSTM better describes the real signal in terms of variance. In the correlation measures, the LSTM outperforms the NARX network for the types *Pearson, Spearman* and *Kendall*, in addition to ensuring the relationship that describes each coefficient by means of the significance value in order of 10^{-55} and statistical power of one (1).

Finally, according to the performance indicators, the complexity of the architecture and the experimentation times; it is concluded that preference cannot be given to the LSTM or NARX architecture. Although the LSTM network presents better regression and correlation performance indicators, the NARX network without taking into account the error generated by the three (3) initial values presents an objective function very close to the LSTM network with a difference of only 0.005032. Additionally, the implementation of the LSTM is more complex in its development and computationally expensive in times of experimentation against the NARX network.

References

1. Fama, E.F.: Efficient capital markets: a review of theory and empirical work. J. Finan. **25**(2), 383–417 (1970)

2. Peters, E.E.: Fractal Market Analysis: Applying Chaos Theory to Investment and Economics. John Wiley & Sons, New york (1994)
3. Sierra, K.J., Duarte, J.B., Pérez, J.M.: Comprobación del comportamiento caótico en bolsa de valores de colombia. Rev. Estrategia Organizacional **2**, 41–54 (2013)
4. Duarte, J.B., Sierra, K.J.: Fractales y caos en el mercado bursátil colombiano. In: 18th Congreso Internacional de Contaduría, Administración e Informática, (Ciudad Universitaria México, D.F.), pp. 1–11 (2013)
5. Martín, I.G.: Análisis y predicción de la serie de tiempo del precio externo del café colombiano utilizando redes neuronales artificiales. Universitas Scientiarum **8**, 45–50 (2003)
6. Agudelo Montoya, A.P.: Modelo de red neuronal para la predicción del precio en bolsa de la electricidad. tesis de maestría, Universidad de Antioquia, Medellín, Colombia (2016)
7. Reyes, L.M.: Aplicación del algoritmo adaboost.rt para la predicción del indice colcap y el diseño de un controlador no lineal. Trabajo de grado para optar por el título de Ingeniero Electrónico (2017)
8. Maciado, J.R.: Desempeño de la técnica de redes neuronales artificiales, frente a los modelos de series de tiempo arima-garch en la predicción de los precios de la acción de bancolombia. Trabajo de grado para optar por el título de Ingeniero Financiero (2013)
9. Macchiarulo, A.: Predicting and beating the stock market with machine learning and technical analysis. J. Internet Bank. Commer. **23**(1), 2–22 (2018)
10. Abbas, R.A., Jamal, A., Ashour, M.A.H., Fong, S.L.: Curve fitting predication with artificial neural networks: a comparative analysis. Periodicals Eng. Nat. Sci. **8**(1), 125–132 (2020)
11. Karakoyun, E.S., Cibikdiken, A.O.: Comparison of arima time series model and lstm deep learning algorithm for bitcoin price forecasting. In: The 13th Multidisciplinary Academic Conference in Prague, (Prague), pp. 171–180 (2018)
12. Diaconescu, E.: The use of NARX neural networks to predict chaotic time series. WSEAS Trans. Comput. Res. **3**(3), 182–191 (2008)
13. Matkovskyy, R., Bouraoui, T.: Application of neural networks to short time series composite indexes: evidence from the nonlinear autoregressive with exogenous inputs (NARX) model. J. Quant. Econ. **17**(2), 433–446 (2019)
14. Matkovskyy, R., Bouraoui, T.: Application of neural networks to short time series composite indexes: evidence from the nonlinear autoregressive with exogenous inputs (narx) model. J. Quant. Econ. **17**(2), 433–446 (2018)
15. Han, J., Kim, S., Jang, M., Ri, K.: Using genetic algorithm and narx neural network to forecast daily bitcoin price. Comput. Econ. **56**(2), 337–353 (2019)
16. Kubat, M.: Neural networks: a comprehensive foundation by simon haykin, macmillan, 1994. Knowl. Eng. Rev. **13**(4) 409-412 (1999). isbn: 0-02-352781-7
17. Hochreiter, S., Schmidhuber, J.: Lstm can solve hard long time lag problems. In: Advances in Neural Information Processing Systems, pp. 473–479 (1997)
18. Li, X., Ghiasi, A., Xu, Z., Qu, X.: A piecewise trajectory optimization model for connected automated vehicles: Exact optimization algorithm and queue propagation analysis. Transp. Res. Part B: Methodological **118**, 429–456 (2018)
19. Ravi, S.S., Rosenkrantz, D.J., Tayi, G.K.: Facility dispersion: heuristic and special case algorithms. Oper. Res. **42**(2), 299–310 (2004)
20. Guo, L., Zhao, S., Shen, S., Jiang, C.: Task scheduling optimization in cloud computing based on heuristic algorithm. J. Netw. **7**(3), 547–553 (2012)
21. Lee, K.S., Geem, Z.W., Lee, S.H., Bae, K.W.: The harmony search heuristic algorithm for discrete structural optimization. Eng. Optim. **37**(7), 663–684 (2005)

22. Altan, A., Karasu, S., Bekiros, S.: Digital currency forecasting with chaotic meta-heuristic bio-inspired signal processing techniques. Chaos Solitons Fractals **126**, 325–336 (2019)

23. Yang, Z., Zhang, J., Tang, K., Yao, X., Sanderson, A.C.: An adaptive coevolutionary differential evolution algorithm for large-scale optimizatio. In: 2009 IEEE Congress on Evolutionary Computation, (Trondheim), pp. 102–109 (2009)

24. Dorigo, M., Di Caro, G.: Ant colony optimization: a new meta-heuristic. In: Proceedings of the 1999 Congress on Evolutionary Computation-CEC99 (Cat. No. 99TH8406), (Washington), pp. 1470–1477, Julio (1999)

25. Bangert, P.: Optimization for Industrial Problems, 1st edn. Springer, Berlin (2011)

26. Poli, R., Kennedy, J., Blackwell, T.: Particle swarm optimization: an overview. Swarm Intell. **1**(1), 33–57 (2007)

27. Huang, C., Li, Y., Yao, X.: A survey of automatic parameter tuning methods for metaheuristics. IEEE Trans. Evol. Comput. **24**(2), 201–216 (2019)

28. Eiben, A.E., Smit, S.K.: Parameter tuning for configuring and analyzing evolutionary algorithms. Swarm Evol. Comput. **1**(1), 19–31 (2011)

29. Mirjalili, S., Jangir, P., Saremi, S.: Multi-objective ant lion optimizer: a multi-objective optimization algorithm for solving engineering problems. Appl. Intell. **46**(1), 79–95 (2017)

30. Mcmahon, D., Topa, D.M.: A Beginner's Guide to Mathematica, New York, pp. 1–6 (2006)

31. Adhikari, R., Agrawal, R.K.: An introductory study on time series modeling and forecasting. arXiv preprint. arXiv:1302.6613 (2013)

32. Peng, Y., Albuquerque, P.H.M., Kimura, H., Saavedra, C.A.P.B.: Feature selection and deep neural networks for stock price direction forecasting using technical analysis indicators. Mach. Learn. Appl. **5**, 100060 (2021)

33. Panigrahi, S., Karali, Y., Behera, H.: Normalize time series and forecast using evolutionary neural network. Int. J. Eng. Res. Technol. **2**, 2518–2522 (2013)

34. Nabipour, M., Nayyeri, P., Jabani, H., Mosavi, A., Salwana, E., Shahab, S.: Deep learning for stock market prediction. Entropy, **2**(8) (2020)

35. Herrera Cofre, D.F., et al.: Predicción para el mercado de acciones con redes neuronales lstm (2020)

36. Singh, A.: Predicting the stock market using machine learning and deep learning. Electron. Res. J. Eng. Comput. Appl. Sci. **2**, 202–219 (2018)

Author Index

A. D. Orjuela-Cañón et al. (Eds.): ColCACI 2022, CCIS 1746, p. 129, 2023.
https://doi.org/10.1007/978-3-031-29783-0

Printed in the United States
by Baker & Taylor Publisher Services